An Introduction to Visual Research Methods in Tourism

"This collation is a triumph of applied thought. Commencing with a tempered coverage of the ontological concerns and the epistemological significances which underpin methodological choice today, the book comprises a foray into many emergent sorts of visual data-mongering. Its best feature is its richness in provision of annotated (easy-to-digest) practical guidelines and chapter summaries. Rakić and Chambers have done a splendid interdisciplinary job."

Professor Keith Hollinshead, University of Bedfordshire, UK

"This volume introduces an array of visual research methods –from using secondary materials to generating visual data and more– relevant to various tourism research projects. I would recommend this book for those who understand the important role of visual information and seek to apply and develop innovative approaches for research in tourism."

Assistant Professor Iis Tussyadiah, Temple University, USA

This is the first book to present, discuss and promote the use of a range of visual methods in tourism studies. It introduces methods ranging from the collection of secondary visual materials for the purposes of analysis (such as postcards, tourism brochures, and websites) and the creation of visuals in the context of primary research (such as photography, video and drawings), to the production of data through photo-elicitation techniques. The book promotes thoroughly underpinned interdisciplinary visual tourism research and includes an exploration of many key philosophical, methodological and (inter)disciplinary approaches. Comprised of five parts: Introduction; Paradigms, Academic Disciplines and Theory; Methods; Analysis and Representation; and Conclusion, this volume informs and inspires its readers through a reliance on theory, examples from tourism studies conducted in various geographical locations and through key pedagogical features such as annotated further readings, practical tips boxes and concise chapter summaries.

This book will be of interest to experienced visual tourism researchers as well as students and scholars who are contemplating the incorporation of visual methods in their studies of Tourism.

Tijana Rakić holds a Ph.D. from Edinburgh Napier University, UK, where she is currently a Lecturer in Tourism and Events and Deputy Postgraduate Tourism Programmes Leader.

Donna Chambers holds a Ph.D. Tourism from Brunel University and currently lectures in tourism and is Programme Director for the M.Sc. International Events Management at the University of Surrey, UK.

Contemporary Geographies of Leisure, Tourism and Mobility

Series Editor: C. Michael Hall

Professor at the Department of Management, College of Business and Economics, University of Canterbury, Private Bag 4800, Christchurch, New Zealand.

The aim of this series is to explore and communicate the intersections and relationships between leisure, tourism and human mobility within the social sciences.

It will incorporate both traditional and new perspectives on leisure and tourism from contemporary geography, e.g. notions of identity, representation and culture, while also providing for perspectives from cognate areas such as anthropology, cultural studies, gastronomy and food studies, marketing, policy studies and political economy, regional and urban planning, and sociology, within the development of an integrated field of leisure and tourism studies.

Also, increasingly, tourism and leisure are regarded as steps in a continuum of human mobility. Inclusion of mobility in the series offers the prospect to examine the relationship between tourism and migration, the sojourner, educational travel, and second home and retirement travel phenomena.

The series comprises two strands.

Contemporary Geographies of Leisure, Tourism and Mobility aims to address the needs of students and academics, and the titles will be published in hardback and paperback. Titles include:

1. **The Moralisation of Tourism**
   Sun, sand....and saving the world?
   Jim Butcher
2. **The Ethics of Tourism Development**
   Mick Smith and Rosaleen Duffy
3. **Tourism in the Caribbean**
   Trends, development, prospects
   Edited by David Timothy Duval
4. **Qualitative Research in Tourism**
   Ontologies, epistemologies and methodologies
   Edited by Jenny Phillimore and Lisa Goodson
5. **The Media and the Tourist Imagination**
   Converging cultures
   Edited by David Crouch, Rhona Jackson and Felix Thompson
6. **Tourism and Global Environmental Change**
   Ecological, social, economic and political interrelationships
   Edited by Stefan Gössling and C. Michael Hall
7. **Cultural Heritage of Tourism in the Developing World**
   Edited by Dallen J. Timothy and Gyan Nyaupane

8. **Understanding and Managing Tourism Impacts**
   An integrated approach
   Michael Hall and Alan Lew

9. **An Introduction to Visual Research Methods in Tourism**
   Edited by Tijana Rakić and Donna Chambers

Forthcoming:

Tourism and Climate Change
Impacts, adaptation and mitigation
C. Michael Hall, Stefan Gössling and Daniel Scott

Routledge Studies in Contemporary Geographies of Leisure, Tourism and Mobility is a forum for innovative new research intended for research students and academics, and the titles will be available in hardback only. Titles include:

1. **Living with Tourism**
   Negotiating identities in a Turkish village
   Hazel Tucker

2. **Tourism, Diasporas and Space**
   Edited by Tim Coles and Dallen J. Timothy

3. **Tourism and Postcolonialism**
   Contested discourses, identities and representations
   Edited by C. Michael Hall and Hazel Tucker

4. **Tourism, Religion and Spiritual Journeys**
   Edited by Dallen J. Timothy and Daniel H. Olsen

5. **China's Outbound Tourism**
   Wolfgang Georg Arlt

6. **Tourism, Power and Space**
   Edited by Andrew Church and Tim Coles

7. **Tourism, Ethnic Diversity and the City**
   Edited by Jan Rath

8. **Ecotourism, NGO's and Development**
   A critical analysis
   Jim Butcher

9. **Tourism and the Consumption of Wildlife**
   Hunting, shooting and sport fishing
   Edited by Brent Lovelock

10. **Tourism, Creativity and Development**
    Edited by Greg Richards and Julie Wilson

11. **Tourism at the Grassroots**
    Villagers and visitors in the Asia-Pacific
    Edited by John Connell and Barbara Rugendyke

12. **Tourism and Innovation**
    Michael Hall and Allan Williams

13. **World Tourism Cities**
    Developing tourism off the beaten track
    Edited by Robert Maitland and Peter Newman

14. **Tourism and National Parks**
    International perspectives on development, histories and change
    Edited by Warwick Frost and C. Michael Hall

15. **Tourism, Performance and the Everyday**
Consuming the Orient
Michael Haldrup and Jonas Larsen

16. **Tourism and Change in Polar Regions**
Climate, environments and experiences
Edited by C. Michael Hall and Jarkko Saarinen

17. **Fieldwork in Tourism**
Methods, issues and reflections
Edited by C. Michael Hall

18. **Tourism and India**
A critical introduction
Kevin Hannam and Anya Diekmann

19. **Political Economy of Tourism**
A critical perspective
Edited by Jan Mosedale

20. **Volunteer Tourism**
Theoretical frameworks and practical applications
Edited by Angela Benson

21. **The Study of Tourism**
Past trends and future directions
Richard Sharpley

22. **Children's and Families' Holiday Experience**
Neil Carr

23. **Tourism and National Identity**
An international perspective
Edited by Elspeth Frew and Leanne White

24. **Tourism and Agriculture**
New geographies of consumption, production and rural restructuring
Edited by Rebecca Torres and Janet Momsen

25. **Tourism in China**
Policy and development since 1949
David Airey and King Chong

26. **Real Tourism**
Practice, care, and politics in contemporary travel culture
Edited by Claudio Minca and Tim Oaks

27. **Last Chance Tourism**
Adapting tourism opportunities in a changing world
Edited by Raynald Harvey Lemelin, Jackie Dawson and Emma Stewart

28. **Tourism and Animal Ethics**
David A. Fennell

Forthcoming:

Gender and Tourism
Social, cultural and spatial perspectives
Cara Atchinson

Liminal Landscapes
Travel, experience and spaces in-between
Edited by Hazel Andrews and Les Roberts

Tourism and War
Edited by Richard Butler and Wantanee Suntikul

Tourism in Brazil
Environment, anagement and segments
Edited by Gui Lohmann and Dianne Dredge

Sexuality, Women and Tourism
Susan Frohlick

Actor Network Theory and Tourism
Ontologies, methodologies and performances
Edited by René van der Duim, Gunnar Thór Jóhannesson and Carina Ren

Backpacker Tourism and Economic Development in the Less Developed World
Mark Hampton

An Introduction to Visual Research Methods in Tourism

Edited by
Tijana Rakić and
Donna Chambers

LONDON AND NEW YORK

First published 2012
by Routledge
2 Park Square, Milton Park, Abingdon, Oxon OX14 4RN

Simultaneously published in the USA and Canada
by Routledge
711 Third Avenue, New York, NY 10017

Routledge is an imprint of the Taylor & Francis Group, an informa business

British Library Cataloguing in Publication Data
A catalogue record for this book is available from the British Library

Library of Congress Cataloging in Publication Data
An introduction to visual research methods/edited by Tijana Rakić and Donna Chambers.
p. cm.
Includes bibliographical references and index.
1. Tourism – Social aspects. 2. Tourism – Psychological aspects. 3. Visual communication – Research. 4. Visual perception – Research. 5. Advertising Tourism. I. Rakić, Tijana. II. Chambers, Donna.
G155.A1I589 2011
910.72—dc22
2011004387

ISBN: 978–0–415–57004–6 (hbk)
ISBN: 978–0–415–57005–3 (pbk)
ISBN: 978–0–203–85586–7 (ebk)

Typeset in Times New Roman
by Florence Production Ltd, Stoodleigh, Devon

Contents

Illustrations

Figures

Tables

Practical Tips

Contributors

Erika Andersson Cederholm is a Senior Lecturer at the Department of Service Management, Lund University. She holds a Ph.D. in Sociology and her research interests embrace the intersection between culture, economy and social interactions, particularly in tourism and hospitality contexts.

William Cannon Hunter Ph.D. is a Professor in the College of Hotel and Tourism Management, Kyung Hee University, Seoul, Korea. His research focuses on destination imagery and problems related to cultural and touristic representations and he is interested in visual research methodologies.

Donna Chambers holds a Ph.D. in Tourism from Brunel University and currently lectures in tourism and is Programme Director for the MSc International Events Management at the University of Surrey. Her research interests and publications include heritage representation in tourism and critical approaches to tourism research.

Michael Haldrup Ph.D. is lecturing in human geography and technology and design studies at Roskilde University, Denmark. He has authored several papers on mobility, tourism, and experience design and co-authored *Performing Tourist Places* (2004, with Bærenholdt, Larsen and Urry) and *Tourism, Performance and the Everyday* (2010, with Larsen).

Salla Jokela is a Ph.D. candidate at the Department of Geosciences and Geography at the University of Helsinki. She is preparing her dissertation in an Academy of Finland funded research project 'Landscape, Icons, and Images'.

Jonas Larsen Ph.D. is a Lecturer in Geography at Roskilde University, Denmark. He is interested in mobility, tourism and media and has published many refereed articles in international journals, book chapters and co-authored four books.

Alison McIntosh Ph.D. is Professor of Tourism at the University of Waikato. Her main research interests are in tourists' experiences of heritage and culture, the subjective, personal and spiritual nature of tourism and hospitality experiences, and qualitative and critical approaches to tourism research.

Naomi Pocock is a Ph.D. candidate at the University of Waikato. Her research interests include the return phase of the travel experience, interpersonal relations in tourism, post-disciplinarity, and innovative approaches to research. Her Ph.D. explores concepts of 'home' for returnees from long-term travel.

Pauliina Raento Ph.D. is Professor of Human Geography at the University of Helsinki and Research Director for The Finnish Foundation for Gaming Research.

Tijana Rakić holds a Ph.D. from Edinburgh Napier University where she is currently a Lecturer in Tourism and Events and Deputy Postgraduate Tourism Programmes Leader. Her research interests and publications lie in the areas of visual research, ethnographic filmmaking, tourism, world heritage and national identity.

Joy Sather-Wagstaff Ph.D. is an Assistant Professor of Anthropology at North Dakota State University. Her research focuses on tourists' experiences in memorial museums and commemorative landscapes, material and intangible heritage, community history/heritage project development, and tourist photography.

Caroline Scarles Ph.D. is a Senior Lecturer in Tourism at the University of Surrey. Her research focuses on the visualities of the tourist experience and sustainable tourism. She is a founding member of the International Network for Visual Studies in Organisations.

Richard Tresidder Ph.D. is a Senior Lecturer in Marketing at Sheffield Business School, Sheffield Hallam University. He has a keen interest in the semiotics of tourism and hospitality and has been working on the development of how semiotics underpin the marketing of destinations, experiences and attractions.

Anne Zahra Ph.D. is a Senior Lecturer at the University of Waikato. Her research interests are tourism research methodologies, the ontological and epistemological foundations of tourism and hospitality research, tourism policy, volunteer tourism and hospitality and society.

Acknowledgements

We would like to thank the chapter authors who shared our passion for the use of visual methods in tourism and who supported our vision to produce the first introductory text dedicated to the subject. Their valued contributions and their patience during the production process are much appreciated.

Part 1

Introduction

1 Introducing visual methods to tourism studies

Tijana Rakić and Donna Chambers

Introduction

An increasing focus on the visual, visuality and the use of visual methods has been evident across a wide range of disciplines and fields of study for quite some time now (e.g., Banks 2001, 2007; Banks and Morphy 1997; Crang 2003; Deveraux and Hillman 1995; Edmonds 1974; Elkins 2008; Emmison and Smith 2000; Hall 1997; Harper 1989, 2002, 2003, 2005; Hill 2008; Hockings 2003; Loizos 1993; Manghani *et al.* 2002; Mirzoeff 1999, 2002; Mitchell 1994; Pauwels 2000, 2004; Pink 2001, 2006; Pink *et al.* 2004; Prosser 1998; Prosser *et al.* 2008; Prosser and Loxley 2008; Rogoff 2000; Rose 2007; van Leuween and Jewitt 2001; Taylor 1994; Stanczak 2007; Wiles *et al.* 2008). In the context of tourism research a significantly greater focus on the visual has also been seen (e.g., Crouch and Lübbren 2003; Crouch *et al.* 2005; Crang 1997; Urry 2002; Selwyn 1996; Jaworski and Pritchard 2005), although discussions surrounding visual research methods have only just recently surfaced (e.g., Feighey 2003; Rakić and Chambers 2009; Rakić and Chambers 2010). Nonetheless, there is a growing recognition of the merits of visual methods within the study of tourism which has, arguably, resulted both from the increasing legitimisation of qualitative research, and the willingness of tourism researchers to explore innovative approaches to research.

Despite the recent popularity of visual methods in tourism research, until the publication of this volume there was no textbook-length publication that could have been used as a reference point by tourism students, early career researchers and established academics. As a result, tourism researchers wishing to employ visual methods needed to refer exclusively to methodological publications outside their field of study. While referring to texts outside one's field of study or discipline as well as experimenting with new methods and approaches to research can undoubtedly be advantageous and rewarding (particularly given the fact that in this process researchers often expand their horizons and create new knowledge), without access to a single reference text linking visual methodologies and their development in the wider social sciences and humanities to potential modes of application in the studies of tourism, the task of locating the relevant literature and developing new, thoroughly underpinned visual approaches to tourism research might prove to be challenging. A number of visual studies in tourism, some of which are remarkably interesting (whether relying on analysis of visual materials collected from secondary sources, on the creation of visual data in the field or elicitation techniques), inevitably seemed to be somewhat experimental – with some researchers failing to appropriately acknowledge their philosophical position, [inter]disciplinary location, and the influence these might have exerted on their choices of visual method[s], techniques of analysis and representation of research findings.

In an attempt to promote thoroughly underpinned visual tourism research while at the same time providing [inter]disciplinary bridges which researchers are encouraged to cross, this volume includes a discussion of issues ranging from questions surrounding the main philosophical approaches, [inter]disciplinary location, research quality criteria, a range of different types of visual methods and different data analysis types to recommendations for potential approaches in publishing visual research findings. Since this is an introductory text, what it does not seek to do is to explore all the answers to all the questions surrounding visual methods in tourism studies. Instead, as this chapter and the remainder of the volume reveals, this text focuses on certain key issues which we and the contributors believe are of importance for tourism scholars who are already using visual methods, for those who are wishing to incorporate visual methods for the very first time as well as for students on undergraduate, postgraduate or doctoral level courses who are perhaps also contemplating the incorporation of visual methods in their studies of tourism.

Visual methods and tourism research

Why should tourism researchers and students seek to incorporate visual methods in their studies? In a recent publication, we argued that there are many reasons for which visual methods are particularly relevant for the studies of tourism (Rakić and Chambers 2010). We placed particular emphasis on the fact that 'much of tourism is about images' and therefore visual methods can play a central role in allowing researchers to access and create knowledge about phenomena which cannot be as readily accessed with the sole use of the more traditional non-visual methods (ibid.). We also emphasised that a subsequent inclusion of visual data in publications can allow tourism researchers to convey findings which cannot be as easily conveyed with the sole use of text, graphs and numbers, as well as that some types of visual research outputs can also enable researchers to share some of the knowledge about tourism with audiences beyond the academic world (ibid.). In addition, given that tourism is a complex field of studies informed by a wide range of disciplines, paradigms and methodological approaches (see Chapters 2 and 3 of this volume) and that tourism can only be studied if 'disciplinary boundaries are crossed' (Graburn and Jafari 1991 in Holden 2005: 1), visual methods might also be perceived as being particularly relevant for tourism research precisely because these can effectively be used within different [inter]disciplinary, philosophical and methodological approaches.

While a wide range of different visual methods have been used in tourism, a broad division between three different types of visual methods can be made in that visuals can be: (1) collected from secondary sources and later studied by relying on analyses such as content or semiotic analysis; (2) created for the purposes of a research project by either the researchers or their research participants; or (3) used to create data by using techniques of elicitation.

Namely, tourism researchers can for example collect and study previously published [secondary] visual data such as postcards, stamps, travel photographs and videos posted on the Internet and social networking sites, images of destinations, tourists, locals, cultural and natural attractions included in guidebooks, promotional campaigns, brochures, travel blogs, travel supplements, television programmes, films and artwork to name just a few examples. A great number of remarkably interesting tourism studies which have analysed visuals which researchers have collected from secondary sources are regularly published (e.g., see Pritchard and Morgan 2003; Tussyadiah and Fesenmaier 2009). In addition to studying visuals from secondary sources, tourism researchers can also, as a part of their primary research, decide to create visual data themselves in the form of researcher-created photography, film, video or drawings which can later be analysed as data, used as illustrations, or to create [audio]visual research outputs such as photo-essays, short videos or documentaries. In this context of researcher-created visual data, it is interesting to note that there are many tourism researchers who are already doing visual research without realising it (see also Kroeber 1994 in Banks and Morphy 1997: 4 for a similar comment for anthropologists), by for example taking photographs while on fieldwork or drawing maps of visitor movements at a popular visitor attraction, some of which visual data they later use as [visual] fieldwork notes, to present some aspects of their research to their colleagues and students, or as illustrations in their publications. What many of these tourism researchers still do not realise is that such researcher-created visuals can be used, not only as illustrations, but also as legitimate 'data' which, in case this fits the research aims and objectives, can also play a more central role in their projects (e.g., Larsen 2005; Rakić 2010). In a similar vein, research participants can also be the ones to create primary visual data such as photography, video and drawings that are later used for analysis and representation of research findings (e.g., Garrod 2008; Son 2005; MacKay and Couldwell 2004). Finally, visuals created by researchers, research participants or those collected from secondary sources can also be used to elicit data in the context of an interview by using elicitation techniques such as photo-elicitation (e.g., see Jenkins 1999; Andersson Cederholm 2004).

Thus, the potential of visual methods to contribute to the creation of new [visual and textual] knowledges about tourism is undoubtedly immense. That said, while visual methods will be suitable for a very wide variety of tourism research projects, and some studies will even necessitate their use (such as for example the studies of postcards or the studies of photographing practices of tourists at a particular place), there will also be projects for which visual methods, or a particular visual method, will not be suitable. Sarah Pink discussed this in the context of visual ethnography, where having arrived on site she felt that it was not appropriate to use researcher-created video as she had originally proposed to do and decided to use photography instead (2001).

Therefore, although applicable in a very wide range of tourism studies, there will also be studies within which it will not be possible to incorporate a particular visual method due to the nature of the particular research topic or context. In fact, as this volume and particularly Part 3 will demonstrate, visual methods are rarely used independently of other methods such as for example interviews and observation in the context of visual ethnographic fieldwork (e.g., Pink 2001) and analysis of both text and images in the context of semiotic analyses of secondary materials such as promotional materials or guidebooks (e.g., Rakić and Chambers 2007). Even if in the early stages of a research project tourism researchers propose to incorporate a particular visual method which seems ideal for its use but which in the latter stages is discovered to be not as ideal as originally believed, in most cases it will still be possible to conduct the research by relying on a different visual research method or on the remainder of the proposed [non-visual] methods.

The book and its chapters

This volume aims to present, discuss and promote the use of a wide range of visual methods and data, including still images (such as photographs, postcards, drawings, postage stamps) and moving images (such as video) within the context of tourism research. In this regard it is informed by perspectives from a range of disciplines including anthropology, geography, psychology and sociology and fields of study such as film, media and visual studies. The volume is therefore interdisciplinary in nature, something which is reflected not only in its content but also in the range of [inter]disciplinary backgrounds of the contributors to this volume. This being the case, it might also prove to be a valuable reference point for studies beyond tourism.

In its recognition that not only is tourism informed by a wide range of disciplines and fields of study and that visual methods can be incorporated within different philosophical, disciplinary and methodological approaches, following this introductory chapter in Part 1 written by the editors, this book includes the much needed Part 2, *Paradigms, academic disciplines and theory*. Its two chapters, each of which is written by one of the editors, provide a robust philosophical and interdisciplinary discussion and serve as an underpinning for the remainder of the volume. Even though discussing what can often be complex questions of philosophy and [inter]disciplinarity, these chapters are written in an uncomplicated style which makes them accessible to a wide readership.

Chapter 2, *Philosophies of the visual [method] in tourism research*, written by Tijana Rakić, engages with the main philosophical issues which would ideally need to be considered in the context of visual tourism research. The chapter contains both an explication of some of the key philosophical positions or paradigms such as positivism, postpositivism, critical theory and constructivism as well as a discussion of debates surrounding the 'realistic' attributes of visuals, the blurred ontological and epistemological boundaries and the modes in which these influence and inform different ways of thinking about, and incorporating, visual methods in tourism research. An overview of the main modes in which constructivism informed the visual aspects of her doctoral research about the Acropolis in Athens serves to elucidate the theoretical discussions in the chapter.

Chapter 3, *The [in]discipline of visual tourism research*, written by Donna Chambers, builds on the philosophical and methodological discussion from Chapter 2 and turns its focus on to epistemological issues of knowledge creation in visual tourism research. Through an exegesis of journal articles published in the mainstream tourism journals, she explores the breadth of disciplinary, theoretical and methodological approaches that have been drawn on in the creation of visual tourism knowledge, demonstrating the need for a continued diversity of disciplinary and methodological approaches. In so doing, she provides a solid underpinning for her discussion of the strategies, tools and different research quality criteria tourism researchers and students might wish to rely on in their visual research projects.

The third part of the book, *Methods* is written by contributors and includes case studies from different countries that guarantee both the geographical scope and relevance of the examples as well as illustrating the diversity of visual methods in tourism studies. Importantly, this is the longest part of the book which explores both the many secondary sources from which visual materials can be obtained for the purposes of research, as well as the methods by which visual data might be produced or used in the context of primary research.

Salla Jokela and Pauliina Raento address the methods of collecting visuals from secondary sources and the dilemmas researchers might face in this process in Chapter 4, *Collecting visual materials from secondary sources.* Jokela and Raento rely on widely applicable and very relevant examples of brochures, postcards, postage stamps, the Internet and historical landscape photographs from their own research in Finland to provide creative ideas and guidance on how secondary visual data can be accessed, collected and organised as well as how choices of visual materials can be justified within a visual research project.

Chapter 5, *Eliciting embodied knowledge and response: respondent-led photography and visual autoethnography*, written by Caroline Scarles, demonstrates how visuals, such as photographs, can be used in creative, reflexive and imaginative ways to produce research data by employing primary research methods of respondent-led photography and visual autoethnography. Scarles relies on both references to seminal theories and existing research surrounding the visual in tourism as well as examples from her own research in Peru to demonstrate the importance of accessing and creating embodied knowedges about tourism as well as the modes in which visuals such as photographs can be used in this process.

In Chapter 6, *Photo-elicitation and the construction of tourist experience: photographs as mediators in interviews*, Erika Andersson Cederholm builds on some of the themes of the previous chapter and discusses the photo-elicitation method further. In addition to contextualising the technique of photo-elicitation as used in sociology and anthropology, Andersson Cederholm incorporates pertinent examples from her own research, namely her studies of returning backpackers and small-scale rural hospitality and tourism businesses in Sweden, and in so doing provides vital guidance to researchers and students who might wish to incorporate this technique in their own studies of tourism.

Chapter 7, *Video diary methodology and tourist experience research*, written by Naomi Pocock, Alison McIntosh and Anne Zahra, moves on to discuss the participant-driven video diarising method in tourism research. While situating this method in the wider traditions of ethnographic filmmaking and other visual approaches to research, Pocock *et al.* emphasise that the use of video in tourism research is still relatively infrequent. In aiming to promote the use of video and provide guidance, they draw on examples from a case study where participant-created video diaries were one of the methods used to study the notion of 'home' for returnees from long-term travel in New Zealand.

William Cannon Hunter introduces the drawing method and its potential in creating knowledge about tourism in Chapter 8, *The drawing methodology in tourism research*. Like other chapters, this chapter also first situates the method and the logic of drawing within disciplines and fields of study such as anthropology, sociology, psychology and the arts before discussing its potential in the studies of tourism. Through extensive and very valuable examples as well as guidance rooted in his own research conducted in Jeju, South Korea, in this chapter Hunter demonstrates the value of the drawing method for tourism projects which aim to access and create knowledge about visual perceptions of research participants.

Part 4, *Analysis and representation* comprises three chapters that discuss the analysis of visual data as well as potential approaches to representing visual data in academic publications.

Chapter 9, *Readings of tourist photographs*, written by Michael Haldrup and Jonas Larsen, explores different types of representational and non-representational readings of photographs, and simultaneously links these to different paradigms, theoretical and methodological approaches to thinking about, and analysing, visual data. In explicating both the theoretical frameworks and providing relevant examples, the authors guide the readers through the research processes and analytical decisions of their studies in Denmark, and elucidate 'how different conceptual lenses produce very different readings

. . . [and] induce significant differences in how we conceive of tourism, tourists and the places of tourism'.

In Chapter 10, *Beyond content: thematic, discourse-centred qualitative methods for analysing visual data*, Joy Sather-Wagstaff, focuses on analysis of visual data by providing an overview of different types of analysis and placing a particular emphasis on the value of qualitative data analyses (QDA) in generating rich insights about the phenomena under study. Sather-Wagstaff relies on relevant examples from her 'research on tourists and tourism at the former site of the World Trade Center (WTC) in New York' to explicate the stages of thematic discourse-centred analysis of visual and textual data, and includes an interesting discussion about the advantages and disadvantages of manual and computer-assisted qualitative data analysis.

In Chapter 11, *Representing visual data in tourism studies publications*, Richard Tresidder engages with the dilemmas tourism researchers and students might face when attempting to publish visual data within the traditionally textual research outputs. In addition to exploring some of the barriers to the publication of visual data, this chapter also provides practical advice and guidance for overcoming these barriers through an ethical and responsible approach to both research design and representation of research findings within which researchers rely on sound academic, legal and ethical principles which underpin both their visual research projects and subsequent publications of research findings.

In conclusion, Part 5 contains Chapter 12, *The future of visual research methods in tourism studies*, written by Donna Chambers and Tijana Rakić, who provide some insights into the future of visual methods in the studies of tourism through a cogent summary of the various chapters in the volume.

Ethics of visual tourism research

While, perhaps surprisingly, none of the chapters in this book focus solely on the ethics of visual research in tourism, there are a number of reasons for this being the case. One of the reasons is that, given that there are a myriad of different ways in which excellent (and ethical) visual research can be undertaken, it would have been an impossible task to attempt to chart the ethical considerations researchers might need to make in the context of their projects within the very limited space of one chapter. Another reason is that with the exception of several very informative and thought-provoking recent publications (e.g., Wiles *et al.* 2008; Prosser *et al.* 2008) and the sections surrounding ethical issues in visual research included in other publications (e.g., Banks 2006; Pink 2001), visual research ethics are not as well charted as word and number based research ethics (Prosser *et al.* 2008). A final reason is that visual research ethics are, similar to wider research ethics, 'contested, dynamic and contextual' (ibid.: 2). This being the case, the contributors to this volume have reflected on some of the key ethical considerations they made in the course of their research projects and, wherever applicable, have provided further guidance. This has been done with a view to motivate tourism researchers to become more aware and engaged with the wider debates surrounding the ethics of visual research, to thoroughly think through their context specific visual research projects, legal and institutional requirements as well as their own moral principles (ibid. 2008) and in so doing foresee and adequately address any potential ethical issues.

It is important, however, to emphasise in this introductory chapter that the ethical considerations visual researchers might need to make are likely to be slightly different to those that need to be considered by more traditional non-visual researchers, especially given that visuals such as photography or video allow for little or no anonymity of the people depicted (Rakić 2010) and that ethical considerations will be dependent not only on the visual method in question but also on the specific research context. In fact, Prosser *et al.* (2008: 2) claim that 'visual methods, and the data they produce, challenge some of the ethical practices of word and number based research, in particular around informed consent, anonymity and confidentiality, and dissemination strategies'.

In this context, researchers and students are advised to thoroughly engage with the existing literature and debates surrounding the ethics of visual research (some of the literature is listed in the annotated further reading section of this chapter and some of the debates are discussed in the chapters included in this volume), ethical guidance, relevant procedures, codes of conduct as well as advice available within their institutions, and ensure that they have adequately considered and attended to any ethical issues which might be relevant for their respective visual research project[s].

Reading the volume

This volume can be read both selectively and as a whole, although those readers who are new to visual methods in tourism would particularly benefit from reading the volume as a whole given that the structure and the content of the book are designed in such a way as to provide a thorough introduction to the areas which we and the contributors thought were of particular philosophical, [inter]disciplinary, theoretical, methodological, analytical, subject-specific and practical relevance.

This volume, as its readers will undoubtedly notice, also includes a number of pedagogical features. In addition to cross-references to previous visual methods publications (often beyond the field of tourism) made in the body of the text, annotated further reading sections are also provided at the end of each chapter. These are intended to serve as a guide to other quality publications in the area of visual methods which the author[s] of each chapter think will be valuable for students, researchers and academics who might not be otherwise aware of these publications or who have not as yet realised their potential application to a particular aspect of visual tourism research. Further on, Part 3, *Methods*, also contains 'Practical Tips' boxes which are intended to assist researchers and students in the avoidance of some of the pitfalls associated with the method discussed in each of these chapters. Finally, in addition to most of the chapters containing examples from research in varying geographical locations which demonstrate how a particular theory, method or approach discussed in the chapter could be applied in practice, all the chapters also feature summaries which are intended to succinctly recap and reinforce the key points its readers are expected to take away from the chapter.

CHAPTER SUMMARY

- Visual methods have been used extensively across various disciplines and fields of study, and their growing popularity in tourism research is also evident.
- There are many excellent, ethical, creative and innovative ways in which visual research in tourism can be undertaken and this volume inevitably introduces only some of the ways in which visual methods have been used in the studies of tourism to date.
- Visual methods are rarely used independently of other non-visual methods.
- Visuals can be collected from secondary sources, created by the researchers or their research participants, or used to create data by relying on elicitation techniques.
- Since the potential of visual methods in the creation of new knowledges about tourism is undoubtedly immense, this volume aims to encourage and inspire both researchers and students to be innovative and creative by engaging in thoroughly underpinned, ethical, interdisciplinary visual research approaches to studying and thinking about tourism.

Annotated further reading

Banks, M. (2001) *Visual Methods in Social Research*, London: Sage.

In addition to excellent guidance for the use of various types of visual methods within a number of academic disciplines, this book also contains a very informative summary of the potential as well as the history of visual methods in the social sciences.

Prosser, J., Clark, A., and Wiles, R. (2008) 'Visual Research Ethics at the Crossroads', retrieved 25 May 2010 from www.eprints.ncrm.ac.uk/535/1/10–2008–11-realities-prosseretal.pdf.

This is another excellent paper which discusses not only the current state of affairs of visual research ethics and its futures, but also its complex ethical and moral decision-making processes and the four key factors which inform these decisions – namely law and copyright; regulations, frameworks and ethical committees; issues of confidentiality and anonymity; and researchers' own moral principles.

Prosser, J., and Loxley, A. (2008) 'Introducing Visual Methods', retrieved 15 November 2009 from www.ncrm.ac.uk/research/outputs/publications/methods review/MethodsReviewPaperNCRM-010.pdf.

This very informative introductory visual methods review paper is likely to prove to be of relevance for both those who have used visual methods in the past and those who are planning to embark on their visual research journey for the first time.

Rose, G. (2007) *Visual Methodologies: An Introduction to the Interpretation of Visual Materials*, 2nd edn, London: Sage.

This is an excellent choice of a reference text for researchers and students who are seeking further guidance in their choices, and subsequent analysis, of visual materials as it covers a number of different types of interpretation such as semiology, psychoanalysis, content and discourse analysis, audience studies, as well as anthropological approaches to interpretation.

Wiles, R., Prosser, J., Bagnoli, A., Clark, A., Davies, K., Holland, S., *et al.* (2008) 'Visual Ethics: Ethical Issues in Visual Research', retrieved 22 November 2009 from www.eprints.ncrm.ac.uk/421/1/MethodsReviewPaperNCRM-011.pdf.

This paper is an excellent introductory review of some of the key ethical issues that visual researchers often need to consider when incorporating the use of visuals such as photographs, films and video. It is likely to be a valuable resource for both researchers in the stages of contemplating visual research and those whose current visual research projects are underway.

References

Andersson Cederholm, E. (2004) 'The Use of Photo-elicitation in Tourism Research: Framing the Backpacker Experience', *Scandinavian Journal of Hospitality and Tourism*, 4: 225–41.

Banks, M. (2001) *Visual Methods in Social Research*, London: Sage.

Banks, M. (2007) *Using Visual Data in Qualitative Research*, London: Sage.

Banks, M., and Morphy, H. (eds) (1997) *Rethinking Visual Anthropology*, Wiltshire: Yale University Press.

Crang, M. (2003) 'The Hair in the Gate: Visuality and Geographical Knowledge', *Antipode*, 35: 238–43.

Crang, M. (1997) 'Picturing Practices: Research through the Tourist Gaze', *Progress in Human Geography*, 21: 359–73.

Crouch, D., Jackson, R., and Thompson, F. (eds) (2005) *The Media and the Tourist Imagination: Convergent Cultures*, London: Routledge.

Crouch, D., and Lübbren, N. (eds) (2003) *Visual Culture and Tourism*. Oxford: Berg.

Devereaux, L., and Hillman, R. (eds) (1995) *Fields of Vision: Essays in Film Studies, Visual Anthropology and Photography*, Berkley: University of California Press.

Edmonds, R. (1974) *About Documentary: Anthropology on Film: A philosophy of people and art*, Dayton: Pflaum Publishing.

Elkins, J. (2008) *Visual Literacy*, New York: Routledge.

Emmison, M., and Smith, P. (2000) *Researching the Visual*, London: Sage.

Feighey, W. (2003) 'Negative Image? Developing the Visual in Tourism Research', *Current Issues in Tourism*, 6: 76–85.

Garrod, B. (2008) 'Exploring Place Perception: A Photo-based Analysis', *Annals of Tourism Research*, 35: 381–401.

Grady, J. (2008) 'Visual Research at the Crossroads', *Forum: Qualitative Social Research*, 9: retrieved 15 May 2010 from www.qualitative-research.net/index.php/fqs/article/view/1173/2619.

Hall, S. (ed.) (1997) *Representation: Cultural Representation and Signifying Practices*, London: Sage.

Harper, D. (1989) 'Visual Sociology: Expanding the Sociological Vision', in G. Blank, J. McCartney and E. Brent (eds), *New Technologies in Sociology: Practical Applications in Research and Work*, New Jersey: Transaction Publishers.

Harper, D. (2002) 'Talking About Pictures: A Case for Photo-elicitation', *Visual Studies*, 17: 13–26.

Harper, D. (2003) 'Reimagining Visual Methods: Galileo to Neuromancer', in N. K. Denzin and Y. S. Lincoln (eds), *Collecting and Interpreting Qualitative Materials*, 2nd edn, London: Sage.

Harper, D. (2005) 'What's New Visually?', in N. K. Denzin and Y. S. Lincoln (eds), *The SAGE Handbook of Qualitative Research*, 3rd edn, London: Sage.

Hill, A. (2008) 'Writing the Visual', retrieved 26 April 2010 from www.cresc.ac.uk/publications/documents/wp51.pdf.

Hockings, P. (ed.) (2003) *Principles of Visual Anthropology*, 3rd edn, Berlin and New York: Mouton de Gruyter.

Holden, A. (2005) *Tourism Studies and the Social Sciences*. London: Routledge.

Jaworski, A., and Pritchard, A. (eds) (2005) *Discourse, Communication and Tourism*, Clevedon: Channel View Publications.

Jenkins, O. H. (1999) 'Understanding and Measuring Tourist Destination Images', *International Journal of Tourism Research*, 1: 1–15.

Larsen, J. (2005) 'Families Seen Photographing: Performativity of Tourist Photography', *Space and Culture*, 8: 416–34.

Loizos, P. (1993) *Innovation in Ethnographic Film: From Innocence to Self-consciousness 1955–1985*, Manchester: Manchester University Press.

MacKay, K. J., and Couldwell, C. M. (2004) 'Using Visitor-employed Photography to Investigate Destination Image', *Journal of Travel Research*, 42: 390–96.

Manghani, S., Piper, A., and Simons, J. (2002) *Images: A Reader*, London: Sage.

Mirzoeff, N. (1999) *An Introduction to Visual Culture*, London: Routledge.

Mirzoeff, N. (ed.) (2002) *The Visual Culture Reader*, 2nd edn, London: Routledge.

Mitchell, W. J. T. (1994) *Picture Theory: Essays on Verbal and Visual Representation*, Chicago: University of Chicago.

Pauwels, L. (2000) 'Taking the Visual Turn in Research and Scholarly Communication', *Visual Sociology*, 15: 7–14.

Pauwels, L. (2004) 'Filmed Science in Search of a Form: Contested Discourses in and Sociological Filmmaking', *New Cinemas*, 2: 41–60.

Pink, S. (2001) *Doing Visual Ethnography*, London: Sage.

Pink, S. (2006) *The Future of Visual Anthropology: Engaging the Senses*, London: Routledge.

Pink, S., Kürti, L., and Afonso, A. L. (eds) (2004) *Working Images: Visual Research and Representation in Ethnography*, London: Routledge.

Pritchard, A., and Morgan, M. (2003) 'Mythic Geographies of Representation and Identity: Contemporary Postcards of Wales', *Journal of Tourism and Cultural Change*, 1: 111–30.

Prosser, J. (ed.) (1998) *Image-based Research: A Sourcebook for Qualitative Researchers*, London: Falmer Press.

Prosser, J., and Loxley, A. (2008) 'Introducing Visual Methods', retrieved 15 November 2009 from www.ncrm.ac.uk/research/outputs/publications/methodsreview/MethodsReviewPaperNCRM-010.pdf.

Prosser, J., Clark, A., and Wiles, R. (2008) 'Visual Research Ethics at the Crossroads', retrieved 25 May 2010, from www.eprints.ncrm.ac.uk/535/1/10-2008-11-realities-prosseretal.pdf.

Rakić, T. (2010) 'Tales from the Field: Video and its Potential in Creating Cultural Tourism Knowledge', in G. Richards and W. Munsters (eds), *New Perspectives on Cultural Tourism Research*, Wallingford: CABI.

Rakić, T., and Chambers, D. (2007) 'World Heritage: Exploring the Tension Between the National and the 'Universal'', *Journal of Heritage Tourism*, 2: 145–55.

Rakić, T., and Chambers, D. (2009) 'Researcher with a Movie Camera: Visual Ethnography in the Field', *Current Issues in Tourism*, 12: 255–70.

Rakić, T., and Chambers, D. (2010) 'Innovative Techniques in Tourism Research: An Exploration of Visual Methods and Academic Filmmaking', *International Journal of Tourism Research*, 12: 379–89.

Rogoff, I. (2000) *Terra Infirma: Geography's Visual Culture*, London: Routledge.

Rose, G. (2007) *Visual Methodologies: An Introduction to the Interpretation of Visual Materials*, 2nd edn, London: Sage.

Son, A. (2005) 'The Measurement of Tourist Destination Image: Applying a Sketch Map Technique', *International Journal of Tourism Research*, 7: 279–94.

van Leeuwen, T., and Jewitt, C. (eds) (2001) *Handbook of Visual Analysis*, London: Sage.

Selwyn, T. (ed.) (1996) *The Tourist Image: Myths and Myth Making in Tourism*, London: Wiley.

Stanczak, G. C. (ed.) (2007) *Visual Research Methods: Image, Society and Representation*, London: Sage.

Taylor, L. (ed.) (1994) *Visualising Theory*, London: Routledge.

Tussyadiah, I., and Fesenmaier, D. (2009) 'Mediating Tourist Experiences: Access to Places via Shared Videos', *Annals of Tourism Research*, 36: 24–40.

Urry, J. (2002) *The Tourist Gaze*, 2nd edn, London: Sage.

Wiles, R., Prosser, J., Bagnoli, A., Clark, A., Davies, K., Holland, S., *et al.* (2008). 'Visual Ethics: Ethical Issues in Visual Research' retrieved 22 November 2009 from www.eprints.ncrm.ac.uk/421/1/MethodsReviewPaperNCRM-011.pdf.

Part 2

Paradigms, academic disciplines and theory

2 Philosophies of the visual [method] in tourism research

Tijana Rakić

Introduction

The fragmentation of tourism studies and theory as well as the range of disciplines and philosophical approaches that inform the knowledge created within the study of tourism have been discussed widely (e.g., see Echtner and Jamal 1997; Franklin and Crang 2001; Tribe 1997; Tribe 2005). At the same time, tourism researchers have also been criticised for insufficient awareness and subsequent recognition of their philosophical positions or paradigms (e.g., see Phillimore and Goodson 2004b; Tribe 2006; Coles *et al.* 2009) as well as for their 'unwillingness to reach across disciplinary and methodological boundaries' (Echtner and Jamal 1997: 86).

Jennings (2001) interestingly noted that most tourism research textbooks do not include a discussion surrounding the key philosophical approaches or paradigms, despite the fact that these inevitably influence and underpin the research process. While this phenomenon might be understandable given not only that the studies of tourism are informed by a wide range of disciplines and philosophical approaches but are also a relatively recent academic endeavour (Echtner and Jamal 1997; Tribe 1997; Phillimore and Goodson 2004b) that have only lately started showing signs of maturity (Ayikoru 2009; Phillimore and Goodson 2004b), this state of affairs surely cannot be constructive for the overall advancement of tourism studies. Marking, to my mind, a positive trend in this context is the publication of a range of recent volumes in tourism which provide guidance and rich discussions, as well as encouraging thoroughly underpinned and innovative approaches to research (e.g., see Phillimore and Goodson 2004a; Ateljevic et al 2007; Tribe 2009). Although these have not necessarily focused on innovative methods, a number of tourism research textbooks also include a discussion (e.g., see Jennings 2001; Long 2007) or, at the very least, a mention of paradigms (e.g., see Ritchie *et al.* 2005; Veal 2006; Finn *et al.* 2000). In attempting to contribute to this trend further as well as provide a philosophical backdrop for the remainder of this volume, this chapter engages with some of the main philosophical considerations that would ideally need to be made in the context of [visual] tourism research.

The first part of this chapter contains a discussion surrounding some of the key philosophical positions or paradigms namely positivism, postpositivism, critical theory and constructivism and the modes in which these might influence and inform the use of a range of visual methods in tourism. The later discussions, which focus on blurred boundaries between the different approaches and particularly between ontological realism and relativism, argue that, especially within qualitative [visual] tourism research projects, constructivism with its relativistic ontology and subjectivist epistemology might be the most viable philosophical position. Simultaneously, however, within these discussions I do not deny the legitimacy of other paradigms. Therefore, while discussing a range of different paradigms, I do not attempt to recommend a single 'correct' paradigm for visual tourism research projects. Instead, throughout the chapter, and particularly within the discussion of examples from a recent visual research project of my own, I aim to draw attention to the importance of avoiding 'philosophical amateurism' in visual research practices in tourism.

Philosophical positions or paradigms and the studies of tourism

In the context of academic research and philosophy of science, paradigm, a concept rooted in Thomas Kuhn's (1962) *Structure of Scientific Revolutions*,[1] is probably best understood as a 'basic belief system' (Guba 1990a: 9) and a way of seeing and thinking about the world (Long 2007) within a disciplined inquiry (Guba 1990b). Although an agreement on a finite number of paradigms which guide academic research practices does not exist (given that paradigms as well as the blurred philosophical positions develop and evolve over time), there seems to be substantial agreement about four key paradigms or world views – positivism, postpositivism, critical theory and constructivism (ibid.). These and other philosophical positions are, in the main, characterised by answers to three crucial questions (ibid.: 18, italics in original):

- *Ontological*: What is the nature of the 'knowable'? Or, what is the nature of reality?
- *Epistemological*: What is the nature of the relationship between the knower (the inquirer) and the known (or knowable)?
- *Methodological*: How should the inquirer go about finding out knowledge?

Different paradigms provide different sets of answers to these three questions and given that these are philosophical positions, the answers contained within different paradigms are ideally not to be perceived as 'absolutely correct' answers. Indeed, had there been only one set of 'correct' answers to these philosophical questions there would be no need whatsoever for the plurality of paradigms (Guba 1990b) that exist not only in the field of tourism studies but across the social sciences and humanities. In fact, different philosophical positions will appeal to, or be developed within, different disciplines, fields of study and communities of researchers while some paradigms will necessarily be more prominent (or some might even argue more appropriate) than others within particular disciplines or topics of research.

Positivism

Within a positivist paradigm, which along with the postpositivist paradigm dominated much of tourism research to date (see also discussions in Ayikoru 2009; Riley and Love 2000; Franklin and Crang 2001; Westwood 2005), answers to ontological questions surrounding the nature of reality and the nature of the knowable usually contain a belief that 'there exists a single reality *out there*, driven by immutable natural laws' (Guba 1990b: 19, italics in original). This ontological position is also known as naïve realism. This position of naïve realism tends to extend to a belief that phenomena studied are both predictable and controllable, that these can be explained by causal relationships, and that findings are 'true' and can be generalised (Guba 1990b; Jennings 2001). In terms of epistemology, or the beliefs related to the relationship of the knower and the known or knowable, inspired by their ontological position, positivists tend to believe that this relationship should be objective or unbiased and value free as well as that objective accounts of the discovered 'true' nature of reality or the knowable can be provided (Guba 1990b; Jennings 2001; Denzin and Lincoln 2005). Subsequently, positivist methodology tends to be marked by deductively created hypotheses or research questions which are then often tested through the utilisation of purely quantitative methods employed within highly controlled conditions of empirical research (Jennings 2001). As an extension of the belief that the knowledge created within positivist research projects is objective and value-free, representations of research findings are usually highly unlikely to contain a textual or visual reference to the researcher(s) who undertook the project. In keeping with the epistemological belief of objectivity, academic texts written by positivists tend to be written in third person narrative.

Postpositivism

Given that it is an adaptation of positivism (Guba, 1990b) postpositivism shares a number of similarities with positivism. In terms of ontological questions, answers provided from a postpositivist position still contain a realist belief that there is a single reality or a single knowable. However, a critical element is added in that postpositivists believe that social reality or the knowable can only be partly perceived or discovered (Guba 1990b; Denzin and Lincoln 2005). Therefore, in terms of their ontological position postpositivists tend to subscribe to critical, rather than naïve realism. Similar to their ontological position, postpositivists also adapted the positivist epistemological position and tend to believe that objectivity in research, although not fully attainable, should still be something that is aimed towards (Guba 1990b). In attempting to achieve as objective and as realistic research findings as possible, the deductive style of postpositivist methodology is therefore usually marked by the utilisation of a mixture of quantitative and qualitative methods, triangulation and significant attempts to base the findings on the widest possible range of information sources (ibid.). Given that postpositivists have departed from naïve realism and added a critical element to their realist ontological beliefs, the representations of findings might, although in very rare occasions, contain a reference to the researchers who undertook the project, and academic texts still tend to be written in a neutral third person narrative.

Critical theory

Critical theory, which Guba also describes as 'ideologically oriented inquiry' (ibid.: 23), significantly departs from positivism and postpositivism in answering some of the key philosophical questions mentioned at the beginning of this section. Although a belief that a single reality or a single knowable exists is shared with positivists and postpositivists, critical theorists also tend to believe that this reality is virtual and that it can only be perceived through the prism of subjective values (Guba, 1990b; Riley and Love 2000; Guba and Lincoln 2005). Therefore, the ontological position of critical theorists is critical realism, although in a different sense to postpositivists, given that critical theorists also tend to extend their critical realist position to historical and value mediated realism (Guba 1990b; Riley and Love 2000; Guba and Lincoln 2005). The subjectivist epistemology to which critical theorists subscribe is inherent to their ontological position and implies that the relationship between the knower and the known is believed to be subjective and value mediated (Guba 1990b). A critical theory-informed methodology is usually also linked to a desire to transform the world, challenge the taken-for-granted beliefs and raise consciousness through reliance on predominantly qualitative methods (Guba 1990b; Riley and Love 2000; Guba and Lincoln 2005; Chambers 2007). Guided by their subjectivist epistemology and their critical, historical and value-laden ontology, within their reflexive research practices, critical theorists are likely both to include textual, and in cases of visual research also visual references to themselves as researchers. Therefore, wherever possible, critical theorists are also likely to write their academic texts in first person narrative.

Constructivism

Constructivism, which is the last of the four key paradigms discussed in this chapter, departs entirely from positivism and postpositivism since it is marked by a completely different set of ontological and epistemological beliefs.

Constructivists are ontological relativists and epistemological subjectivists. In other words, constructivists tend to believe that social reality is plural, created in the minds of individuals and this being the case, they also tend to believe that subjective interactions between the researcher and the researched which lead to co-created findings are the only processes through which knowledge can be created (Guba 1990b; Guba and Lincoln 2005). Supported by the use of qualitative methods, constructivist methodology usually contains hermeneutic and dialectic elements which aid in reconstruction of multiple, subjective realities of their own and of their research participants as well as in the creation of one or more equally valid constructions of knowledge usually reached through consensus (e.g., see discussions in Guba 1990b; Guba and Lincoln 2005; Rakić and Chambers 2009; Rakić 2010). Given that constructivists are, similar to critical theorists, epistemological subjectivists who in the main rely on qualitative methods and subscribe to reflexive research practices, constructivist representations of research findings are also likely to contain textual and visual references to, and discussions about, themselves as researchers within a research project. Led by their epistemological position, similar to critical theorists, constructivists are also likely to write their academic texts in first person narrative.

Philosophies of the visual [method] in tourism research

As mentioned earlier, tourism research has historically been, and continues to be, informed and influenced by a wide variety of disciplines and fields of study which include but are not limited to: economics, anthropology, geography, sociology, psychology, cultural studies, business, marketing, management, and organisational studies, and, as some of the content of this volume demonstrates, even film and visual studies. However, one of the main challenges might not be the multi-, inter-, cross- or even post-disciplinary nature of tourism studies (which might also be perceived as an advantage), but that some paradigms can at times also be taken for granted (Long 2007) within particular areas of tourism research. This being the case, it is important to bear in mind that familiarity with a number of different paradigms is of paramount importance for the overall quality of tourism research projects, and that different [communities of] tourism researchers are likely to favour different paradigms.

Echtner and Jamal (1997) note that within the sociology of tourism there are competing schools of thought that subscribe to either positivist or interpretive approaches. Within geographical studies of tourism, on the other hand, they claim that positivism, which is more prominent within physical geography, has been under fierce attack by human geographers whose contemporary studies of tourism are mostly informed by the more interpretive, subjective, relativist and critical approaches (ibid.). Similar to the situation in human geography, according to Echtner and Jamal, anthropological studies of tourism are also mostly informed by interpretive approaches, while within organisational studies and strategy as well as marketing and consumer studies in tourism, conflicts between positivist and interpretive approaches seem to be evident.

That said, while the focus or the [inter]disciplinary location of a particular tourism related study might provide an indication of the dominant or the most widely acceptable paradigm within the particular area of tourism research, it will not necessarily dictate the philosophical approach to be adopted. As Hollinshead (2004: 75) states while referring to the work of Guba (1990b), Rorty (1979) and Skrtic (1990):

> there is no single set of independent criteria available for all social science, which a neutral observer (if such a person existed!) could ever use to determine which was the best or the most appropriate paradigm for each and every social scientist to work within.

Visual research methods, similar to a number of other methods are only tools within a disciplined enquiry and, irrespective of the paradigm adopted, tourism researchers will have a choice between sourcing visual 'data' from secondary sources or creating visuals as a part of their primary research. Namely, visual [and textual] 'data' from materials such as brochures, guidebooks, press travel supplements, promotional materials, TV travel programmes, videos, postcards and stamps could be sourced from secondary sources and later analysed or interpreted (e.g., see Chapter 4 of this volume). In addition, visuals such as drawings, photographs, video and film can also be created as a part of primary research by either research participants or the researcher(s) undertaking the project (e.g., see Part 3 of this volume). However, the modes in which visual methods might, or might not be, successfully incorporated within a research project will also be dependent on their fit within a particular research topic and context, the remainder of methods, researcher(s) skill-set(s) (Rakić 2010) as well as the extent to which the implications of the underpinning paradigm have been adequately considered.

A number of publications, some of which have been mentioned in this chapter, make a particular distinction between positivist and interpretive approaches. Namely, along with ontological differences, significant differences in epistemological positions of positivist and interpretive approaches tend to result in remarkably different choices of methodologies, methods, types of analysis, as well as styles of representations of knowledge created within a research project (see Table 2.1). For example, in an earlier publication, where among other issues, I had discussed the potential implications of the four key paradigms for researcher-created video (Rakić 2010), I wrote that, informed by their ontological and epistemological positions, researchers who work within positivist approaches are more likely to rely on creating video or filmic

Table 2.1 Key paradigms and potential implications for visual methods

Positivism	*Postpositivism*	*Critical theory*	*Constructivism*
Ontology *(the nature of reality)*			
[naïve] realism: the realit***y*** is driven by unchanging natural laws, it can be discovered	[critical] realism: the realit***y*** is driven by unchanging natural laws, can only be partly perceived/discovered	[critical/historical/value-laden] realism: realit***y*** is virtual, can only be perceived through prism of values	relativism: realit***ies*** are multiple, individually and collectively constructed, local and specific
Epistemology *(relationship of the knower and known)*			
objectivist: researcher is an objective observer, bias excluded, findings considered true	modified objectivist: researchers strive towards a 'regulatory ideal' of objectivity, findings probably true	subjectivist/transactional: values mediate and influence the findings	subjectivist/transactional: findings co-created through interaction
Methodology *(theory and principles of inquiry)*			
hypotheses/research questions, created and tested in controlled environment	modified experimentation, more grounded theory, triangulation, significantly wider range of data	dialogic/transformative/ interactive, seeks to challenge and raise consciousness	hermeneutical/dialectical, reconstruction of realit***ies***, aimed at generating one or more constructions through consensus
Methods *(tools within inquiry)*			
in the main quantitative	along with quantitative, introduces more qualitative	mostly qualitative/interpretation	mostly qualitative/interpretation
Potential implications for visual methods			
visuals tend to be perceived as evidence of reality 'captured' in an image which, used as scientific data, can in turn be used to create objective scientific knowledge; textual or visual references to researcher(s) rarely included		visuals mainly seen as representations of [a] reality as perceived and recorded by the person who created the image in a particular context of time and space; textual and visual references to researcher(s) often included	

Source: Rows 1–4, 'Ontology', 'Epistemology', 'Methodology' and 'Methods' adapted from Guba 1990b; Riley and Love 2000 and Guba and Lincoln 2005; last row 'Potential implications for visual methods' created by the author

footage marked by as little camera movement as possible, create very long takes in which they as researchers are highly unlikely to appear, and later show this footage to their audiences having applied very little or no editing at all. On the other hand, researchers who work within interpretive approaches are more likely to appear in their footage, rely somewhat more on hand-held and moving camera work, shorter takes and editing, as well as to at times rely on author-reflexive narrative when screening their audio-visual work (ibid.).

However, as discussed later on in this chapter, it is important to bear in mind that boundaries between different paradigms can also be blurred, and that there are 'many excellent ways of undertaking visual research and applying visual methods' (Prosser and Loxley 2008: 5). The role of different paradigms in this context can be, among other things, to provide an indication of the modes in which tourism researchers and students might decide to think about the visuals and their position as researchers within their respective research projects. Therefore, despite the fact that the four paradigms are listed in a systematic manner with clearly denoted boundaries, Table 2.1 is not intended to serve as a prescriptive but rather, as an indicative guide for some of the main characteristics of the four paradigms discussed in this chapter and their potential implications for visual methods.

Realism versus relativism and blurred ontological boundaries of the 'visual'

As evident from Table 2.1, in addition to whether visual methods are more likely to be employed as a part of a quantitative, mixed or qualitative primary research method design, one of the additional variations in the treatment of visual methods and data between the different paradigms resides in the differing perceptions of realistic attributes of images. Or, in other words, the extent to which images are perceived to be objective as opposed to subjective representations of reality. Interestingly, images are used within both positivist and interpretive approaches to research (Prosser and Loxley 2008). Some scholars argue that images, referring in particular to photographs, film and video (or other images created through mechanical means), have something inherently realistic and objective about them; others claim that even images such as photographs are created though subjective processes of selection as well as that all images are polysemic in nature in that they carry multiple meanings and lend themselves to multiple interpretations; while some interestingly claim that images are in fact all of these things, objective and subjective, realistic and relative.

Photography provides an exemplary type of image in the context of this discussion of realism versus relativism and the blurred ontological and epistemological boundaries of the visual. In his posthumously published text titled 'The Ontology of the Photographic Image', André Bazin (1960: 7), referring to photography and its differences with painting, for example, writes:

> Originality in photography as distinct from originality of painting lies in the essentially objective nature of photography . . . For the first time an image of the world is formed automatically, without the creative intervention of man. The personality of the photographer enters into proceedings only in his selection of the object to be photographed and by the way of purpose in his mind. Although the final result may reflect something of his personality, this does not play the same role as is played by that of the painter.

While Bazin clearly refers to and attempts to argue that photography is marked by an inherent objectivity since a photograph (unlike a painting) is created through an automated process of a photographic camera, he very interestingly also simultaneously engages in a discussion about the selectivity of the photographing practice, the purpose for which a photograph is made and the fact that, although objective, a photograph (in addition to the photographed external 'reality') may also reflect something about the subjectivity or personality of the photographer. Therefore, given that images of any type, including the ones created with the assistance of automated technological aids such as photographic, digital, video or film cameras, are inevitably selected, framed, edited and made with a specific purpose by an inevitably subjective historically, spatially, culturally and socially situated individual, can images of any kind ever be considered as a purely or exclusively objective and realistic medium? Or would it be an overly simplistic assumption to think of photographs as inherently objective?

In 1965, Pierre Bourdieu (1965: 6) noted that 'even when the production of a picture is entirely delivered over to the automatism of the camera, the taking of the picture is still a choice involving aesthetic and ethical values', while in 2007, Barry M. Goldstein (2007: 645) takes this notion further and suggests that photographs, 'even in the most technically ideal, well intentioned circumstances, can never represent reality' given that 'every image is a result of a large number of technical and aesthetic choices made by the photographer' and that 'each choice introduces subjective elements into the content'. In a similar vein, in 2001, Sarah Pink (2001: 234) poses an interesting question of whether it is in fact possible to observe and record 'reality' and other than suggesting that reality 'is not solely "visible" or observable', she also writes that:

> rather than recording reality on video tape or camera film, the most one can expect is to represent those aspects of experience that are visible [to the person who created the image]. Moreover, these visible elements of experience will be given different meaning as different people use their own subjective knowledge to interpret them.
>
> (ibid.: 24, text in brackets added)

Interestingly, while passionately arguing for a visual ethnographic approach, which seems to be in the main informed by a subjectivist relativist position,

Pink also says that 'in some cases, realist uses of photographic and video images may be appropriate in ethnographic research and representation' (ibid.).

While numerous other arguments and counterarguments vis-à-vis realism and relativism, objectivity and subjectivity, the blurred boundaries and the varying degrees of realism in images can be found in the literature (see for example Banks 1998; Goldstein 2007; Sontag 1977; Eco 1982; Stanzak 2007; Devereaux 1995; Harper 1998; Pink 2001, Wagner 2001; Winston 1998; Prosser and Loxley 2008; Rakić and Chambers 2009), Umberto Eco captivatingly notes that, 'the theory of the photo as an analogue of reality has been abandoned, even by those who once upheld it' (Eco 1982: 33).

Therefore, even though an image produced with the aid of mechanical means such as a photographic camera might be perceived as bearing some connections to 'reality' (e.g., see discussions in Pink 2001) especially given the apparent resemblance of contemporary colour photographs with human visual perception (e.g., see discussions in Goldstein 2007), all images, including the most 'realistic' of photographs, are in fact still subjectively created and later interpreted representations of a subjectively perceived 'reality' at a particular moment in time. This however, does not render photographs or other images as unsuitable 'data' types for academic studies, or as suitable for some philosophical approaches and unsuitable for others, despite the fact that, especially within qualitative [visual] tourism research projects, constructivism with its relativistic ontology and subjectivist epistemology might be perceived as the most viable philosophical position (see also Rakić and Chambers 2009; Rakić 2010).

In fact, visual 'data' have long been used (and are likely to continue being used) in a number of disciplines and fields of study as well as within different philosophical perspectives. No matter the particular context in which tourism researchers and students decide to incorporate images in their research, after having read this and other chapters in this volume, they will ideally be aware that, images are manipulated to a greater or lesser extent at different stages of their production (Goldstein 2007), images are purposefully made by a subjective individual in a particular social, spatial and historic context, and no matter how 'realistic' and 'objective' the content of an image seems to its maker or its viewer, images are inevitably polysemic in nature and are therefore likely not only to be made, but also to be interpreted differently by different people.

While admittedly, I personally favour interpretive over positivist approaches to studying and thinking about both images and tourism, this is not to say that I argue positivist approaches or the positions marked by blurred boundaries between different approaches are illegitimate. In fact, some argue and I also agree that 'such positivist/interpretivist polarity and advocacy is not constructive for the research community' (Prosser and Loxley 2008: 10). That said, there will of course be researchers and research communities who will favour and subscribe to interpretive approaches and those who will subscribe to positivist approaches, and interestingly, some will also find the space of blurred boundaries between ontological realism and relativism as well as between epistemological objectivity and subjectivity to be a fertile ground for their academic work. As Wagner (2001: 7) convincingly argues:

> all science involves a search for relatively reliable ways of identifying, observing, examining and talking about things that at least appear to be objective, external realities; and all science also depends on the subjective experience and knowledge-constructing talents of individual researchers and scientific communities. Ideas about image-based field work can easily get caught up in polarised advocacy of one or the other of these two principles. But photographs, drawings, paintings, video tapes and other cultural artifacts . . . belong to both.

An example of a constructivist approach within a [visual] tourism research project

In attempting to provide an example of some of the modes in which a paradigm can inform a research process, I will now briefly reflect on the modes in which a particular philosophical approach influenced my Ph.D. research. The main focus of this research was the complex relationships between world heritage, tourism and national identity at the Athenian Acropolis (see also Rakić 2008; Rakić and Chambers 2009; Rakić 2010).

Underpinned by a constructivist approach, this interdisciplinary research, informed mainly by [visual] anthropology and [cultural] geography, included a wide variety of textual and visual methods. The visual aspects included semiotic analyses of tourist materials such as postcards, governmental promotional materials and guidebooks that contained a visual and/or textual reference to the Acropolis and a year-long visual ethnographic fieldwork in Athens. During the time of my fieldwork, in addition to collecting some of the tourist materials mentioned above, I also studied the visitation at the Acropolis by relying on audio-visually recorded (filmed and photographed) as well as traditional participant observation, audio-visually recorded interviews, diary-keeping and mapping of visitor movements and favourite spots at the site.

Visual research methods therefore, both in terms of collecting and interpreting visual and textual data from secondary sources (i.e., postcards, governmental promotional materials and guidebooks) and in terms of creating visuals in the field (i.e., photographs, videos and mapping of visitor movements and favourite spots), played a central role in this research project. The question is, how did my constructivist approach inform and influence the research process, from its initial stages, through to fieldwork, analysis and interpretation, to the representation of research findings?

Given that the word strictures of this chapter, and the volume as a whole, do not permit an in-depth reflexive account of the wide variety of modes in which my constructivist position informed this research process, inevitably

I include a summary of some of the key points instead (for a more in-depth account refer to Rakić 2008; Rakić and Chambers 2009; Rakić 2010). First of all, within the early stages of this research, the fact that I was coming from a subjectivist relativist position implied that my research was very likely to be marked by a 'relativist ontology, subjectivist epistemology, hermeneutic methodology and qualitative research methods' (Rakić 2010: 133, see also Table 2.1). It also implied that I was likely to be treating 'reality' 'as relative, personally and collectively constructed (i.e., plural) and "knowledge" as subjective, co-created, and situated' (ibid.). Consequently, this meant that I was not likely to perceive myself according to a positivist tradition, as a person who was looking for 'objective universal truth', but rather that I was most likely to perceive myself as a situated ethnographer central within this 'context specific knowledge creation' (ibid.).

In the fieldwork context I therefore sought to understand the point of view of the visitors to the Acropolis, rather than only observing their activities and behaviour, as well as perceiving the fieldwork as an extension of my own lived experience, and interviews as meaningful dialogues (see also Williamson 2006; Holstein and Gubrium 2008; Rakić and Chambers 2009). This being the case, within filmmaking practices I also relied somewhat more on dynamic filming and hand-held camera than would have been the norm in early ethnographic filmmaking practices (see also Rakić 2010). In the later stages of analysis and creation of research outputs such as the Ph.D. thesis, various journal articles, book chapters, conference papers, an ethnographic film and a two-minute video produced as a part of this project, I inevitably adopted an interpretive approach to analysis as well as a reflexive style in the representation of research findings, in which, wherever possible, I included myself in both the texts (i.e., by writing in first person narrative) and the visuals (i.e., by appearing in the film).

CHAPTER SUMMARY

- Although tourism studies are informed by a wide range of disciplines, fields of study and philosophical approaches, in this chapter I emphasised that tourism researchers are also simultaneously criticised for insufficient awareness of philosophical approaches (e.g., see Phillimore and Goodson 2004b; Tribe 2006; Coles *et al.* 2009), as well as for their 'unwillingness to reach across disciplinary and methodological boundaries' (Echtner and Jamal 1997: 86).
- This being the case, within this chapter I provided a philosophical backdrop for the remainder of this volume and in so doing included an elementary exploration of some of the key philosophical approaches or paradigms such as positivism, postpositivism, critical theory and constructivism as well as discussed the blurred ontological and epistemological boundaries of the visual.

- Visual research methods, which are undoubtedly becoming ever more important, seem to be predominantly used within qualitative or interpretive approaches in the wider social sciences. Despite this being the case, and the fact that, personally, I also favour interpretive over positivist approaches and ways of thinking about both images and tourism, I reiterated that visual methods and data can be effectively incorporated within different philosophical paradigms.
- Finally, by emphasising that different paradigms, or ways of seeing and thinking about the world (Long 2007) within a disciplined inquiry (Guba 1990b), inevitably inform and underpin the entire research process, in this chapter I also explicated some of the reasons why 'philosophical amateurism' should be avoided as well the reasons why a consideration of the answers to the key ontological, epistemological and methodological questions as well as the modes in which these answers inform the visual aspects of research should lie at the heart of any visual research project in tourism.

Note

1 Interestingly, according to Masteman (1970 in Guba 1990b), Kuhn himself used the term paradigm in twenty-one different ways. While this may be perceived as problematic by some, others like Guba (1990b: 17) claim that 'it is important to leave the term in such a problematic limbo, because it is then possible to reshape it as our understanding of its many implications improves'.

Annotated further reading

Goldstein, B. M. (2007) 'All Photos Lie: Images as Data', in G. C. Stanczak (ed.), *Visual Research Methods: Image, Society and Representation*. London: Sage.
This book chapter contains a rich discussion surrounding the extent to which images can be considered as accurate representations of reality. Although the chapter mainly focuses on photographic images, the philosophical discussions surrounding the nature of images as data are also relevant for other types of visual data and this chapter is therefore likely to be of interest to visual tourism researchers.

Guba, E. C. (1990b) 'The Alternative Paradigm Dialog', in E. C. Guba (ed.), *The Paradigm Dialog*. London: Sage.
Although published twenty years ago this book chapter and edited volume are among Guba's most influential publications, which marked some of the later discussions surrounding the four key paradigms. These are likely to be of particular interest to tourism researchers and students who wish to explore some of the ontological, epistemological and methodological positions of these paradigms in greater depth.

Guba, E. C., and Lincoln, Y. S. (2005) 'Paradigmatic Controversies, Contradictions, and Emerging Confluences', in N. K. Denzin and Y. S. Lincoln (eds), *The SAGE Handbook of Qualitative Research*. London: Sage.
This book chapter and edited volume contain the more recent debates surrounding the nature of paradigms and qualitative research across a wide spectrum of academic

disciplines and is likely to be of great interest to tourism scholars and students who wish to engage with these debates further.

Phillimore, J., and Goodson, L. (eds). (2004a) *Qualitative Research in Tourism: Ontologies, Epistemologies and Methodologies*. London: Routledge.
This edited volume contains excellent discussions surrounding the nature of qualitative research, ontologies, epistemologies and methodologies in tourism, as well as discussions surrounding the position of tourism studies within the trends evident in the wider social sciences. This volume is likely to be of interest to researchers and students who subscribe to qualitative approaches in their studies of tourism.

Wagner, J. (2001) 'Does Image-based Field Work Have More to Gain from Extending or Rejecting Scientific Realism? An Essay in Review'. *Visual Sociology*, 16: 7–21.
This journal article contains a very interesting debate surrounding scientific realism and image-based fieldwork. Through a discussion of Cole's (1997) *Doing Documentary Work* and Pink's (2001) *Doing Visual Ethnography*, this article poses an interesting question of whether we should extend or reject scientific realism, a question which is undoubtedly also of relevance to visual tourism researchers.

Winston, B. (1998) '"The Camera Never Lies": The Partiality of Photographic Evidence', in J. Prosser (ed.), *Image-based Research: A Sourcebook for Qualitative Researchers*. London: Falmer Press.
This book chapter once again engages in philosophical discussions surrounding the nature of images and the popular myth that images such as photographs and documentary films can provide unambiguous and complete evidence about the real world. This being the case, this chapter is also likely be of great relevance for visual tourism researchers.

References

Ateljevic, I., Pritchard, A., and Morgan, N. (eds). (2007) *The Critical Turn in Tourism Studies: Innovative Research Methodologies*. Oxford and Amsterdam: Elsevier.

Ayikoru, M. (2009) 'Epistemology, Ontology and Tourism', in J. Tribe (ed.), *Philosophical Issues in Tourism*. Bristol: Channel View Publications.

Banks, M. (1998) 'Visual Anthropology: Image, Object and Interpretation', in J. Prosser (ed.), *Image-based Research: A Sourcebook for Qualitative Researchers*. London: Falmer Press.

Bazin, A. (1960) 'The Ontology of Photographic Image' [Translated by Hugh Gray]. *Film Quarterly*, 13: 4–9.

Bourdieu, P. (1965 [1990]) 'Introduction', in P. Bourdieu with L. Boltanski, R. Castel, J.-C. Chamboredon and D. Schnapper, *Photography: A Middle-brow Art* [translated by Shaun Whiteside]. Stanford: Stanford University Press.

Chambers, D. (2007) 'Interrogating the "Critical" in Critical Approaches in Tourism Research', in I. Ateljevic, Pritchard, A., and Morgan, N. (eds), *The Critical Turn in Tourism Studies: Innovative Research Methodologies*. Oxford: Elsevier.

Coles, T., Hall, C. M., and Duval, D. T. (2009) 'Post-disciplinary Tourism', in J. Tribe (ed.), *Philosophical Issues in Tourism*. Bristol: Channel View Publications.

Denzin, N. K., and Lincoln, Y. S. (2005) 'Introduction', in N. K. Denzin and Y. S. Lincoln (eds), *The SAGE Handbook of Qualitative Research*, 3rd edn. London: Sage.

Devereaux, L. (1995) 'Cultures, Disciplines, Cinemas', in L. Devereaux and R. Hillman (eds), *Fields of Vision: Essays in Film Studies, Visual Anthropology and Photography*. London: University of California Press.

Echtner, C. M., and Jamal, T. (1997) 'The Disciplinary Dilemma of Tourism Studies', *Annals of Tourism Research*, 24: 868–83.

Eco, U. (1982) 'Critique of the Image', in V. Burgin (ed.), *Thinking Photography*. London: Macmillan.

Finn, M., Elliott-White, M., and Walton, M. (2000) *Tourism and Leisure Research Methods: Data Collection, Analysis and Interpretation*. Harlow: Pearson Education Limited.

Franklin, A., and Crang, M. (2001) 'The Trouble with Tourism and Travel Theory?', *Tourist Studies*, 1: 5–22.

Goldstein, B. M. (2007) 'All Photos Lie: Images as Data', in G. C. Stanczak (ed.), *Visual Research Methods: Image, Society and Representation*. London: Sage.

Guba, E. C. (1990a) 'Foreword', in E. C. Guba (ed.), *The Paradigm Dialog*. London: Sage.

Guba, E. C. (1990b) 'The Alternative Paradigm Dialog', in E. C. Guba (ed.), *The Paradigm Dialog*. London: Sage.

Guba, E. C., and Lincoln, Y. S. (2005) 'Paradigmatic Controversies, Contradictions, and Emerging Confluences', in N. K. Denzin and Y. S. Lincoln (eds), *The SAGE Handbook of Qualitative Research*. London: Sage.

Harper, D. (1998) 'An Argument for Visual Sociology', in J. Prosser (ed.), *Image-based Research: A Sourcebook for Qualitative Researchers*. London: Falmer Press.

Hollinshead, K. (2004) 'A Primer in ontological Craft: The Creative Capture of People and Places through Qualitative Research', in J. Phillimore and L. Goodson (eds), *Qualitative Research in Tourism: Ontologies, Epistemologies and Methodologies*. London: Routledge.

Holstein, J. A., and Gubrium, J. F. (2008) 'Constructionist Impulses in Ethnographic Fieldwork', in J. A. Holstein and J. F. Gubrium (eds), *Handbook of Constructionist Research*. New York: Guilford Press.

Jennings, G. (2001) *Tourism Research*, Milton: John Wiley & Sons.

Kuhn, T. S. (1962) *The Structure of Scientific Revolutions*. Chicago: Chicago University Press.

Long, J. (2007) *Researching Leisure, Sport and Tourism*. London: Sage.

Phillimore, J., and Goodson, L. (eds). (2004a) *Qualitative Research in Tourism: Ontologies, Epistemologies and Methodologies*. London: Routledge.

Phillimore, J., and Goodson, L. (2004b) 'Progress in Qualitative Research in Tourism: Epistemology, Ontology and Methodology', in J. Phillimore and L. Goodson (eds), *Qualitative Research in Tourism: Ontologies, Epistemologies and Methodologies*. London: Routledge.

Pink, S. (2001) *Doing Visual Ethnography*. London: Sage.

Prosser, J., and Loxley, A. (2008) 'Introducing Visual Methods', retrieved 15 November from www.ncrm.ac.uk/research/outputs/publications/methodsreview/MethodsReviewPaperNCRM-010.pdf.

Rakić, T. (2008) World Heritage, Tourism and National Identity: A Case Study of the Acropolis in Athens, Greece, Ph.D. Thesis. Edinburgh Napier University, Edinburgh.

Rakić, T. (2010) 'Tales from the Field: Video and its Potential in Creating Cultural Tourism Knowledge', in G. Richards and W. Munsters (eds), *Cultural Tourism Research Methods*. Wallingford: CABI.

Rakić, T., and Chambers, D. (2009) 'Researcher with a Movie Camera: Visual Ethnography in the Field', *Current Issues in Tourism*, 12: 255–70.
Riley, R., and Love, L. L. (2000) 'The State of Qualitative Tourism Research', *Annals of Tourism Research*, 27: 164–87.
Ritchie, B. W., Burns, P., and Palmer, C. (eds). (2005) *Tourism Research Methods: Integrating Theory With Practice*. Wallingford: CABI.
Sontag, S. (1977) *On Photography*. London: Penguin Books.
Stanczak, G. C. (2007) 'Introduction: Images, Methodologies and Generating Social Knowledge', in G. C. Stanczak (ed.), *Visual Research Methods: Image, Society and Representation*, London: Sage.
Tribe, J. (1997) 'The Indiscipline of Tourism', *Annals of Tourism Research*, 24: 638–57.
Tribe, J. (2005) 'New Tourism Research', *Tourism Recreation Research*, 30: 5–8.
Tribe, J. (2006) 'The Truth about Tourism', *Annals of Tourism Research*, 33: 360–81.
Tribe, J. (ed.) (2009) *Philosophical Issues in Tourism*. Bristol: Channel View Publications.
Veal, J. A. (2006) *Research Methods for Leisure and Tourism: A Practical Guide*, 3rd edn. Harlow: Pearson Education Limited.
Wagner, J. (2001) 'Does Image-based Field Work Have More to Gain from Extending or Rejecting Scientific Realism? An Essay in Review', *Visual Sociology*, 16: 7–21.
Westwood, S. (2005) 'Out of the Comfort Zone: Situation, Participation and Narrative Interpretation', in *The First International Congress of Qualitative Inquiry Conference Proceedings*. University of Illinois at Urbana-Champaign.
Williamson, K. (2006) 'Research in Constructivist Frameworks Using Ethnographic Techniques', *Library Trends*, 55: 83–101.
Winston, B. (1998) '"The Camera Never Lies": The Partiality of Photographic Evidence', in J. Prosser (ed.), *Image-based Research: A Sourcebook for Qualitative Researchers*. London: Falmer Press.

3 The [in]discipline of visual tourism research

Donna Chambers

Introduction

This chapter continues the philosophical discussion commenced by Rakić in Chapter 2, but in this chapter I wish to focus specifically on those epistemological issues associated with the use of the visual in tourism research. That is, in this chapter I will seek to address some specific epistemological questions including: how do we know what we know about the visual in tourism research? What are the current boundaries of knowledge about the visual in tourism research? What is legitimate or valid knowledge in visual tourism research? In seeking to address these questions I will discuss the multidisciplinary and interdisciplinary nature of visual tourism research which draws on, and integrates a number of disciplines and theoretical approaches from the social sciences, especially though not exclusively the main disciplines of geography, sociology and psychology, anthropology and their subdisciplines.

The chapters in Part 2 of this text provide examples of how tourism knowledge is created through the use of a variety of visual methods such as photography, video and drawings, but in this chapter I wish to present a critical snapshot of the various disciplinary and theoretical perspectives which have so far informed the use of the visual in tourism research through an exegesis of some of the published works on the visual in the mainstream tourism literature. In so doing I have deliberately omitted any publications by the authors who have contributed to this text for two main reasons. The first is that I believe it is necessary to provide a general overview for the reader of the way in which visuals are used within wider tourism research. The second reason is that the other contributors to this text will 'speak' for themselves throughout the remaining sections of this book.

Importantly, the chapter will also include a commentary on the tools which can be potentially used by students and researchers in order to ensure that their visual tourism studies are rigorous and thus worthy of serious consideration particularly given that studies of tourism are still predominantly located within business and management schools where visual methods have not been widely legitimated.

The [in]discipline of visual tourism research

It will be apparent to avid researchers and students of tourism that the chapter title bears a strong family resemblance to the paper by Tribe (1997) on the [in]discipline of tourism in which he sought to address epistemological concerns in the context of tourism studies in general. In this article Tribe made no mention of those epistemological issues specifically related to the use of visual methods in tourism despite the central role of the visual in tourism. This section of the chapter is consistent with Tribe's discussion in so far as it concludes that in order to create knowledge, researchers and students of tourism should utilise a diversity of disciplinary and theoretical approaches. However it departs from Tribe's work in that it recognises the centrality of the visual in tourism and focuses exclusively on how knowledge is created within this context.

Eight years ago Feighey (2003) argued that in the context of tourism studies anthropological and sociological approaches to visual research were conspicuously absent. In this section I will undertake a review of some of the journal articles that have used visual methods in tourism research, focusing on their disciplinary and theoretical approaches and the variety of methods which they have utilised. The word strictures of this chapter do not allow for the inclusion of an exhaustive number of journal articles and it is recognised that what is provided is only a snapshot of the articles that have utilised visual methods in tourism research. But the aim of the chapter is not comprehension but rather it is hoped that the discussion will illuminate the way in which the visual has been used to create knowledge about tourism. Further the exclusive focus on journal articles is a result of the absence of any books that deal specifically with visual methods in tourism research, a lack that the current text intends to partially address. While there is no book-length work that deals exclusively with visual research methods in tourism, this is not to deny the existence of texts that explore the role of the visual within tourism and the link between visual culture and tourism. Chief among these are works such as that by Crouch and Lubben (2003) on visual culture and tourism; Rojek and Urry's (1997) work titled *Touring Cultures: Transformations of Travel and Theory*; Selwyn's (1996) work titled *The Tourist Image: Myth and Myth Making in Tourism* and of course the seminal text by John Urry (1990) on the tourist gaze.

However while all of these texts are, to one extent or another, about the role of the visual in tourism studies and provide useful insights on how knowledge is created in this context, I felt that it would be appropriate in this chapter to focus exclusively on journal articles as there has not to date been a discussion of the way in which these publications (which are more numerous than books and often more easily accessible to students and researchers) draw on a variety of disciplinary and theoretical approaches and methods to create visual tourism knowledge. In this exegesis of journal articles, I discovered that visual tourism research draws on a multiplicity of disciplinary and theoretical perspectives.[1] Sociological approaches were evident contrary to Feighey's (2003) claim, although there was still a dearth of publications which expressly utilised anthropological approaches.

Importantly also, in my exegesis of tourism journal publications, which included visual research methods, it was apparent that in some instances where the visual was incorporated it was often used as a method of research with little theoretical or methodological contextualisation. Indeed Banks (2001: 2, citing Mead 1995) had indicated that the 'social sciences are disciplines of words in which the visual has been relegated to a supporting role'. In other words, in many of the tourism articles examined where visual methods were utilised, the visual was often not central to the research and there was invariably insufficient discussion about the epistemological and theoretical basis for the use of visual methods. This was particularly evident in those articles with a marketing studies orientation that focused specifically on destination image (these were often underpinned by sociological and psychological insights on the characteristics and sources of consumer image perceptions and consumer behaviour) (cf. Kim and Richardson 2003).

In many of the publications also the use of still photographs and quantitative analytical techniques predominated. Naoi *et al.* (2006), in a study that was underpinned by environmental psychology, used photographs to examine how visitors' mental states affected their classification of the physical and affective features of an historical district in Germany. Using repertory grid analysis and laddering techniques the research revealed the complex nature of visitors' evaluation of place. Fairweather and Swaffield (2001), again drawing on insights from environmental psychology, used photographs of different landscapes to interpret the ways in which Kaikoura, New Zealand, was experienced by visitors to the destination. They indicated that the thirty landscape photographs used in the study were 'surrogates for landscape experiences' (ibid.: 220). The photographs were combined with interviews and the techniques used were the Q method and factor analysis both of which are often used in landscape research where respondents are presented with a number of photographs and asked to sort these, for example, according to their level of importance or significance. Using photographs in this way to measure experiences raises important epistemological questions that will be discussed in the second section of this chapter.

There were also interesting studies which utilised other disciplinary and theoretical approaches. Markwell (1997) drew on social-geographical perspectives to explore the spatial, temporal and social dimensions of photography in order to understand a nature based tour experience. In this study Markwell drew on an eclectic number of ethnographic techniques namely on-site observations, post-tour interviews, photographic collections taken by tour participants (2,680 of which were content-analysed), diaries from some of the participants and the author's own narrative of the tour. From this study, Markwell concluded that:

> The taking of photographs allowed the tour participants considerable power over the way they constructed their tangible memories of the tour . . . very few images of the mundane, the domestic or the unattractive were captured by the cameras thus reinforcing the myth of the perfect (or at least near-perfect holiday) . . . it gives a false impression of the holiday experience as one devoid of domestic activity and routine, aspects of everyday life from which many people are trying to escape.
>
> (Markwell 1997: 153)

MacKay and Smith (2006) adopted a psychological approach in their examination of age related differences in memory of tourism advertising using both text and visual formats. One of the hypotheses that was developed for testing in this study was whether framed (or labelled) pictures of destinations are recalled in the same way by different age groups as are text descriptions of the destinations. Here the authors were informed by the literature on cognitive gerontology, advertising and frame theory. The results of their study revealed that there were no differences between younger and older adults in terms of the information recalled and the number of elaborations of the framed photographs. However, differences became evident when the features of the destination were presented in a written rather than a visual format with the younger adults having greater levels of recall than their older counterparts. The contribution to tourism knowledge here is that it adds to our understanding of the way in which different forms of destination advertising are processed by different age groups and this is important for destination marketing.

As indicated, most of the journal articles reviewed utilised still photography whether they were taken by the respondents themselves (as in Garrod's 2008 study that examined place perception based on volunteer employed photography; Caton and Santos' 2007 study of photographs taken by students on a study-abroad programme) or by the researchers as in landscape studies (cf. Fairweather and Swaffield 2001, mentioned above) or as pictures in tourism brochures (cf. Echtner and Prasad 2003), postcards (cf. Markwick 2001; Sirakaya and Sonmez 2000; Burns 2004) or advertising (cf. MacKay and Smith 2006, detailed above). Importantly as intimated, many of these studies also drew on quantitative techniques such as content and factor analyses.

In a unique study, Tribe (2008) undertook an analysis of a large quantity of artwork (paintings) in order to investigate the knowledge that could be gained about tourism from these paintings. He analysed both the images and the text accompanying these paintings in a method which he termed 'virtual curating' which he explains as one in which the researcher 'assumed the role of a tourism art curator where the output is a display of works organised into viewing galleries supported by an exhibition guide' (ibid.: 926). This method, he claimed represented a 'fusion of elements of grounded theory, content analysis and researcher artistry' (ibid.: 926). While Tribe did not engage with the literature surrounding the use of the visual in general and within tourism research in particular, this example represents another way in which visuals can potentially be used in the creation of knowledge about tourism.

There were also some studies that used moving images such as films or videos to create tourism knowledge. In this regard Yan and Santos (2009) provided a detailed analysis of the first national Chinese promotional video titled *China Forever*. In this study the video was seen as a 'significant and efficient socio-cultural purveyor' (ibid.: 299). The authors drew on post-colonial theories and critical discourse analytic techniques. From this study the authors coined the term 'self orientalism' to suggest that Orientalism is also reinforced, constructed and circulated by the Orient and not just by the West. In this way, they argue, the Orient is itself complicit in its representations of itself. Tussyadiah and Fesenmaier's (2009) study was influenced by the sociology and marketing literature in a study on how Internet videos were used to mediate tourism experiences of destinations. They used a method known as 'netnography' in which they examined all the videos on YouTube that contained touristic activities in New York City. They examined not just the visual content of these videos but also performed a content analysis of the text contained in the reviews of these videos. They concluded that 'travel related videos are shown to be powerful as media of "transportation" within the concept of virtual mobility' (ibid.: 37).

In terms of the use of film there is a defined area of tourism studies on film- or movie-induced tourism in which, among other things, films are analysed in order to create new understandings of tourism. In this context, Frost (2010), drawing on marketing studies (and specifically motivation and image theories), examined twenty-two commercial films about the Australian outback in a longitudinal study spanning over sixty years, in order to discern the messages contained in these films and the experiences they 'promised' for tourists to the outback. He concluded from this study that through the films tourists were 'promised' profound, life-changing experiences.

In one of the few studies underpinned by an anthropological approach, Palmer and Lester (2007) examined the documentary film produced by Dennis O'Rourke titled 'Cannibal Tours' in order to discern how the film-maker used the documentary as a narrative to 'frame' tourism. Using analytic techniques found in film studies and informed by the structuralist and post-structuralist perspectives of noted theorists such as Foucault, Levi-Strauss and Barthes, they concluded that the film 'supports multiple oneiric subjectivities – O'Rourke's, the tourists', the viewers and those of us "reading" the film' (ibid.: 103). Norris Nicholson (2006) undertook a study on the use of home videos to document people's experiences of travel. Using insights from an eclectic mix of disciplines and fields of study including social history, cultural studies, historical geography, identity and memory theories, material culture and emerging literature on archival film, Norris considered the contribution of 'non-professionally made travel related footage material to understanding how 20th century touristic visual practices evolved' (ibid.: 14).

What these brief discussions have demonstrated is the breadth of disciplinary and theoretical perspectives that have been used to inform our understanding of tourism through the use of visuals. However in this context the mainstream tourism literature seems to contain a focus on sociological, psychological and geographical approaches (and to a much lesser extent, anthropological) with the study of destination images and consumer perceptions predominating. This provides support for Stanczak's (2007: 8) argument that 'images need not – in fact, should not-be considered the province of one discipline or held to one set of readings'. The discussions have also demonstrated the variety of visuals which have been utilised although the focus here seems to have been on still or photographic images with a dearth of studies drawing on moving images (the exception here might be those studies on film or movie induced tourism). The discussions so far have also illustrated the many analytic techniques which have been used although there appears to be a focus on quantitative techniques such as content analysis, notwithstanding the fact that there is evidence of the use of more qualitative techniques such as semiotic and discourse analysis.

This snapshot suggests too that the very nature of visual research allows for significant creativity and originality in approaches which will allow us to push the boundaries of knowledge creation about tourism phenomena. Indeed in the examples presented above it was clear that visuals were used to create a plethora of knowledge about tourism phenomena. Some might argue that this [in]discipline of the visual within the context of tourism leads to methodological and theoretical anarchy and it is therefore impossible to establish what is legitimate or valid knowledge in visual tourism research. Indeed, particularly with regard to qualitative visual tourism researchers, it could be argued that there is an irony in our rejection of any kind of data management as a protest against the dominance of traditional forms of empiricism (Wagner 2007). This is an important issue and so the next section will seek to address the question of the legitimacy of visual tourism research.

The legitimacy of visual tourism research

In this section I intend to probe the concepts of validity and reliability and to demonstrate the extent to which these are applicable for the purpose of legitimating visual tourism research. In doing so, it will become apparent to readers that the discussion in this part of the chapter is heavily skewed towards a discussion of the legitimacy of qualitative approaches to visual research which are normally underpinned by subjectivist epistemologies. This is for two principal reasons. First, I feel that the legitimacy of quantitative research, which is underpinned by objectivist epistemologies, and which inaugurated the largely statistical and essentialist notions of validity and reliability, is already well established in social science research. Visual tourism research, which is influenced by positivism and attendant quantitative approaches, therefore draws on these concepts in order to establish their legitimacy. There is still, however, a lack of consensus on how to establish the legitimacy of visual tourism studies that draw on qualitative approaches. Second, and importantly, the bias in this chapter towards qualitative approaches is also a result of my own philosophical perspective which is interpretive and which recognises the subjective and plural nature of knowledge. Indeed, I recognise that it is not just academic communities (or researchers) that create and define knowledge within tourism research. That said, I must however acknowledge, as others in this volume have done and as some of the journal articles previously discussed have demonstrated, that interpretive and qualitative approaches are not the only ways to the creation and construction of knowledge of the visual in tourism research. I also acknowledge that epistemological assumptions about subjective versus objective empiricism present perennial challenges for the social sciences and in the context of visual research there is a continuous shifting, a blurring of boundaries between the two approaches (Stanczak 2007).

Even with this in mind, Wagner (2007) indicates that there is an enduring irony in social science research in so far as research that is removed from its real world origins is 'regarded as more empirically sound' (ibid.: 27) than research which emerges from direct observations in natural settings as is the case in many kinds of visual research in tourism such as photography. Yet there are issues of legitimacy even with visuals such as photographs that are taken in 'real' world settings. For example, if tourists or the members of host societies pose for photographs within the context of a visual research project, does this kind of contrived picture-taking distort the 'reality' of the phenomenon under study? Do cropped and juxtaposed images, made especially possible through digital photography, create false impressions?

In answer to these epistemological questions, Wagner indicates that certain knowledges can be gleaned from what on the surface might appear to be 'unrealistic' representations of the social world, as posed photographs can 'provide valuable evidence of how people want to be seen by others' (ibid.) and 'keeping data as "raw" as possible can also reduce their usefulness in answering empirical questions' (ibid.: 28) that are of concern to the tourism researcher. Interestingly also Goldstein (2007) argues that all photos lie. Specifically he argues that in the case of photographic images:

> Every image is manipulated, thus no image represents reality. Content depends on a large number of technical and aesthetic choices made by the photographer, based on his or her intent.
>
> (Goldstein 2007: 79)

Jacobsen (2007) provides an overview of the use of landscape perception methods in tourism studies that were based on the use of photographs and indicates that the validity of these methods is open to question. This is because these methods invariably rely on the use of still photographs as 'surrogates' of the landscape in question rather than respondents' direct experiences with these landscapes. Indeed Jacobsen suggests that a

> key concern has been with representational validity, the extent to which landscape perceptions, preferences and judgements based on photographs correspond to responses elicited by direct experience with the landscape nominally represented.
>
> (ibid.: 239)

Jacobsen cites Scott and Canter (1997) who have argued that 'a main problem for researching landscape through photographic representation and/or simulation is that humans experience landscapes when they are actually in a location' (Jacobsen 2007: 240). Jacobsen goes on to suggest that on-site studies could potentially address the issue of representational validity that arises through the use of photographs as landscape 'surrogates'.

Yet, it is these ambiguous issues that challenge the extent to which visual tourism research is perceived as rigorous and empirically sound. I argue that as visual tourism researchers, especially those of us working within interpretive paradigms, we need to appreciate that any knowledge that is created through our visual research projects is a reflection of 'reality' only within particular social and cultural settings at particular historical moments and from particular viewpoints. This is in keeping with the constructivist philosophical approach discussed by Rakić in Chapter 2, and implies that visual tourism researchers should not be concerned with absolute notions of truth but should rather be interested in the 'partial and multiple truths of image collections related to a particular project or study' (Wagner 2007: 29). Still, Stanczak (2007: 6) argued that 'an epistemological wariness still challenges the validity of images as data today'.

However, Morse and Barrett *et al.* (2002) rightly suggest that rigour is central to research, as without it research might be perceived as fictitious and devoid of value or utility. This they continued was true for both quantitative and qualitative research and it was wrong for qualitative researchers to reject notions of reliability and validity as these were essential to establishing the legitimacy of research. They make reference especially to the work of Lincoln and Guba (1985) who sought to replace the terms reliability and validity with other concepts that they believed were more appropriate for qualitative research, namely credibility, transferability, consistency and confirmability, each of which will be briefly explained later on in this chapter. However, before proceeding further with the discussion we need to understand what is meant by these twin terms of reliability and validity. Validity refers to the extent to which research findings give the 'correct' answer or are interpreted in a 'correct' way (Kirk and Miller 1986). However, it has been argued that the question of validity relates to a positivist epistemology in which it is believed that there is some 'isomorphism between research findings and reality' (Lincoln and Guba 1985, as cited in Murphy *et al.* 1998: 170) and that the extent of this isomorphism can be ascertained through the use of rigorous statistical techniques. However, the suggestion is that validity is much more complex because it has an internal and external dimension. Internal validity refers to the 'coherence and consistency of a piece of research, and in particular how well the data presented support the researcher's conclusions' (Tonkiss 1998: 259). External validity refers to 'whether the findings are generalizable to other research or social settings' (ibid.). It might be argued that while good visual research can lay claim to an internal validity (that is coherency and consistency with regard to the particular research question/s) the issue of external validity or generalisability is not so readily addressed.

On the other hand reliability refers simply to the extent to which a 'measurement procedure yields the same answer however and whenever it is carried out' (Kirk and Miller 1986:19). Reliability is about whether research results will be consistent and accurate over time. The reliability question for visual tourism researchers is whether it is possible for their research to be reproduced or repeated over time. In other words, will the visual tourism research yield the same or similar results if it is conducted in the same way at a later point in time or by another researcher? In quantitative research Kirk and Miller (ibid.: 41–2) suggest that there are three types of reliability: (1) the degree to which a measurement given repeatedly remains the same; (2) the stability of a measurement over time; and (3) the similarity of measurements within a given time period. From this definition of reliability it is axiomatic that a reliable measure does not imply validity for if one's measurement tool is faulty then while it will yield the same measurements over time this does not mean that these measurements are valid. The opposite however does not hold. That is, it is argued that a valid study is also a reliable study as argued by Lincoln and Guba (1985: 316) who assert that 'since there can be no validity without reliability, a demonstration of the former is sufficient to establish the latter'. However this is perhaps a rather simplistic viewpoint as given the subjectivist epistemology which underpins qualitative research reliability is, prima facie, irrelevant. So one can perhaps unceremoniously dismiss reliability as an objective of visual tourism research that is underpinned by a subjectivist epistemology and instead debates should rather surround the validity or simply the truthfulness of visual tourism research. In other words, why should we believe the interpretations arrived at by visual tourism researchers?

That said, it might be inappropriate to adopt carte blanche the criteria of one epistemological approach to assess the research undertaken using another, generally antithetical perspective. That is, it is inappropriate to speak, for example, of issues of validity when referring to qualitative approaches like some of those adopted in visual tourism research because this fails to reflect the important epistemological and ontological underpinnings of this kind of research. In this regard, Schwandt (1994: 122) has argued that the issue of validity is not applicable to qualitative approaches and suggests instead that 'criteria such as thoroughness, coherence, comprehensiveness, and so forth' should be used. Further, he claims, it should be asked whether 'the interpretation is useful, worthy of adoption and so on'. In addition, it has been argued that the production of 'valid' knowledge is necessarily intertwined with relationships of power. In other words, knowledge in academic research will be perceived to be valid in so far as it corresponds with the dominant epistemologies and ontologies of those invested with the power to make knowledge claims. According to Humberstone (1997: 201):

> Knowledge constituted by research becomes acceptable/unacceptable, valid/invalid depending upon whether it 'fits' with the values, assumptions and ideologies of those in a position to legitimate its credibility.

Still, if issues of validity are inappropriate for interpretive, qualitative research such as that which permeates some of visual research studies in tourism, how then can one justify one's assumptions? Are the interpretations obtained through visual tourism research projects not merely solipsistic accounts? As mentioned previously, Lincoln and Guba (1985) cited in Murphy *et al.* (1998: 170–71) while subscribing to a constructivist perspective, nevertheless attempted to develop a list of criteria which could be used to assess the soundness of qualitative research many of which simply paralleled those established in quantitative research. These criteria included:

- *Credibility*: intended to replace the notion of validity and referring to the extent to which the research subjects concurred with the researcher's findings.
- *Transferability*: to replace the concept of generalisability and referring to the provision by the researcher of 'thick descriptions' of research phenomena which were capable of being utilised in different research settings.
- *Consistency or dependability*: intended to replace the criteria of reliability and referring to what is deemed the 'trackable variance' of sources. That is the extent to which the research design may have induced change in the phenomena under study.
- *Confirmability*: to replace the concept of objectivity and referring to the process by which the researcher has arrived at her/his conclusions. In this regard, they argue that researchers should provide an 'audit trail', which would detail and examine the process by which the researcher conducted his/her research and arrived at his/her conclusions.

Guba and Lincoln thus recommended certain strategies that could be used to achieve these criteria such as peer debriefing, prolonged engagement and persistent observation, audit trails and member checks. The characteristics of the researcher such as responsiveness, adaptability and sensitivity were also important (Guba and Lincoln 1981, as cited in Morse and Barrett *et al.* 2002).

However, the establishment of these criteria by Lincoln and Guba was severely criticised because it was felt that 'criteria of warrantability' (Murphy *et al.* 1998: 172) were specious in the context of a constructivist perspective underpinned by an epistemological subjectivism and an ontological relativism. That is, it was believed that the establishment of criteria are incongruous in the context of a philosophical perspective in which it is believed that truth is perspectival and therefore that there are multiple truths all of which can lay claims to 'truthfulness.' Indeed the 'reflexive epistemologies of much of visual research hold that the meaning of the images resides most significantly in the ways that participants interpret those images, rather than some inherent property of the images themselves' (Stanczak 2007: 11).

The criticisms of Lincoln and Guba's attempt to develop criteria for establishing the soundness of qualitative research led to them formulating even more complex criteria although they did not abandon their constructivist perspective (see Murphy *et al.* 1998: 172). Still it is my belief that Guba and Lincoln, through their various attempts to establish 'check lists' for the evaluation of qualitative research in a manner which simply paralleled those existing for quantitative research failed to adequately appreciate the implications of their own constructivist methodological assumptions. In this context, in the tourism literature, Jamal and Hollinshead (2001) have emphasised that what is important for qualitative research is the ability to gauge the results of the research against its own objectives rather than against some external reality. The qualitative visual researcher must be open to the idea that there might be multiple interpretations of his/her results and should recognise the possibility of different or even competing interpretations. The qualitative visual researcher must also be reflexive where his/her own assumptions are questioned. It must be reiterated that the importance of adopting qualitative techniques within the context of visual tourism research lie in their 'interpretive commitment to processes of meaning in social life . . . and an approach to knowledge which sees this as open rather than closed' (Tonkiss 1998: 260). The thesis of Howarth (2000), while it was concerned with discourse analysis as a qualitative analytical technique is nevertheless instructive for visual researchers in tourism. He explains that the 'validity' of discourse analysis (and I would here include also visual analysis), can be determined by the extent to which they provide plausible new interpretations of the phenomena they set out to investigate:

> the empirical accounts which [visual researchers] produce have to be evaluated as particular interpretations of the research objects they have constructed, and not as confirming or refuting instances of a separately constituted empirical theory. The ultimate tribunal of this evaluation is the community of scholars who judge the interpretations proffered, and the adequacy or inadequacy of [visual methodology] as a whole depends on its ability to engender plausible accounts of social phenomena. In this

> sense, the ultimate criterion for judging the adequacy of [visual methodology] as a whole is pragmatic; it can be evaluated by the degree to which it makes possible new and meaningful interpretations of the social and political phenomena it investigates.
>
> (Howarth 2000: 130, text in brackets added)

This approach might not be fully accepted by those tourism scholars built from the positivist mould and who still exert considerable influence in tourism academia. These have been derogatorily referred to by Hollinshead (1999) as 'quantifrenics' and 'methodolatrists' who believe in the sanctity of statistical measures of validity, reliability and objectivity without accepting that these measures are themselves only social constructions. Indeed, Hollinshead espouses the view that,

> in social science, there are only ever interpretations, whether the given researcher be 'bible-bashing' positivist or 'heretical'/'upstart' bricoleur. There is not anything – any data or subject, anywhere – which is able to *speak for itself*.
>
> (ibid.: 1999: 276, emphasis in original)

For their part Morse and Barrett *et al.* (2002) argued that most of the strategies suggested by Guba and Lincoln could only be achieved *after* the research had been conducted and there existed no strategies to achieve research rigour *during* the process of research. In this context, Morse and Barrett *et al.* argued for a return, by qualitative researchers, to notions of reliability and validity as their dismissal of these notions has led to research which is unreliable and invalid, has made it difficult for qualitative researchers to obtain funding, to get published and to be taken seriously by practitioners and policy makers. They argue that there should be strategies in place during the process of research to verify the validity and reliability of the research. These strategies, which they term verification strategies, should be built into the qualitative research process and include ensuring methodological coherence, sampling sufficiency, developing a dynamic relationship between sampling, data collection and analysis, thinking theoretically, and theory development (Morse and Barrett *et al.* 2002).

While there is some merit in the arguments put forward by Morse and Barrett *et al.*, I still feel that, in case your visual research project is underpinned by one of the interpretive paradigms, there should be ways of legitimating your project that do not rely on positivist legacies such as reliability and validity (the former of which I have already indicated is irrelevant for qualitative visual tourism researchers). What I would say though is that it is certainly vital to ensure that your visual tourism project is academically rigorous and in this regard it is important to first ensure that you have a well thought out research design that outlines the specific issues that you seek to explore and how the study will be organised to address these issues. I argue that this is so, regardless of the kind of data that you choose to examine or the kind of visual research method that you choose to employ. As part of this research design you must ensure that you address issues to do with your own role within the research project, the skills and knowledge that you will bring to bear on the research process and the extent to which you will be able to obtain cooperation and relevant information from the research participants or for those tourism researchers working with secondary data, the extent to which you will be able to access this data. Based on all these discussions it is now possible to outline a potential process that might be used as a 'toolkit' in undertaking visual research in tourism in order to ensure its legitimacy (and these are relevant whatever your philosophical perspective):

1 *Identification of the general area of interest*: This will necessarily be influenced by the philosophical and subjective 'positionality' of the visual tourism researcher.
2 *Preliminary review of relevant literature*: This is necessary in order to determine the nature of the extant research in the general area of interest and to determine whether there are any gaps that could potentially be explored
3 *Definition of research problem*: In this stage the problem to be addressed should be narrowed down into one or more research questions and this should ideally emerge from the preliminary review of the literature or from initial exploratory fieldwork.
4 *Immersion*: This should involve extensive engagement with methodological and theoretical issues to do with the visual in general and with the visual in tourism research in particular.
5 *Conceptualisation*: This should involve an interpretation and understanding of the research question(s) in light of the theoretical and methodological insights gained.
6 *Data collection*: At this stage there needs to be a consideration of the various visual research methods available, their suitability for the research project, and the skills of the visual researcher in terms of the visual method to be selected; the appropriateness of the selected visual method in terms of addressing the research question(s); the accessibility of participants or data to be used for the research. In this stage also you should consider the ethical implications of your research.[2]
7 *Analysis of data and re-conceptualisation*: This stage involves the analysis of the data collected and might involve the use of quantitative and/or qualitative techniques. The analysis of the data might lead to a further clarification of the concepts and arguments.
8 *Reflections*: This stage involves a retrospective analysis of the research process, your own role in the research, its limitations and possible arenas for the extension of the study in a future research project.

It is important to note that while the aforementioned stages suggest a linear, discrete and sequential process, this might not be the case in practice. This is because visual research will often involve a constant iterative process in which concepts and logics and the researcher's own position (and often that of participants) are refined and clarified in the light of new interpretations emerging from the immersion in, and engagement with, the literature and with the data obtained from the visual research project.

Another issue that might concern researchers who use visual methods and which point to the extent of legitimacy of the visual research project, is that of selectivity. For example, in photography or video (whether this be researcher- or participant-created or found from secondary sources), why are some pictures or videos taken and not others? In using secondary visual materials like postcards and brochures, why are some images selected and not others? Indeed in some cases, as has been demonstrated in some of the articles mentioned in this chapter, visual tourism researchers often draw on a variety of sources all in one research project. Readers of this chapter might find such a commingling of evidence 'wantonly eclectic or absurdly disparate' (Lowenthal 1985), but it will be recalled that an important consideration in visual research is not so much comprehensiveness with regard to data sources but rather 'to make cogent what might otherwise go unnoted' (ibid.: xxvi) through verbal or written sources. Indeed, it must be stated that the materials collected for visual tourism research might be neither exhaustive nor statistically systematic. A large number of visual materials might be consulted, but an even larger number might be omitted. However, importantly, for qualitative visual tourism researchers, exhaustion of sources and statistical systematicity in data collection are by no means necessary prerequisites. This is because often qualitative visual tourism research seeks to provide alternative and plausible interpretations rather than the kind of definitive and prescriptive statements found in positivist, quantitative studies. As Said (1995: 4) contended in his seminal study on *Orientalism*, it should be recognised that:

> even with the generous number of books and authors that I examine, there is a much larger number that I simply have had to leave out. My argument, however, depends neither upon an exhaustive catalogue of texts dealing with the Orient nor upon a clearly delimited set of texts, authors, and ideas that together make up the Orientalist canon. I have depended instead upon a different methodological alternative – whose backbone in a sense is the set of historical generalisations.
>
> (ibid.: 1995)

In other words, the aim of research of this kind is not to present a 'whole', 'complete' or essentialist 'truth' about the phenomenon under study but rather to present alternative meanings, and interpretations which might facilitate and generate discussion and further research.

CHAPTER SUMMARY

I would like to conclude this chapter with a brief summary of the important points that should be considered by students and researchers as they seek to undertake visual tourism research projects which add to our knowledge and understanding of the ubiquitous phenomenon of tourism:

- The creation of knowledge in visual tourism research requires an engagement with a variety of disciplinary and theoretical perspectives and methods of research. In this regard visual tourism research has no epistemological boundaries.
- The extent to which the knowledge created through visual tourism research is deemed legitimate or 'truthful' will depend on the extent to which your research project is rigorous. Rigour requires that the research process that you undertake is clearly outlined from the initial point of conceptualisation to the final stage of reflection and this should be so whether your research is based on quantitative or qualitative approaches.
- The [in]discipline of visual research in tourism allows for unlimited creativity and imagination on the part of researchers and students of tourism. It is only thus that new knowledge about tourism can and will emerge.

Notes

1 I recognise that some articles in which visuals are used in tourism research do not appear in mainstream tourism journals. But given that the focus of this text is on tourism as a distinct field of study and given that there are numerous journals on tourism, it is apt that my exegesis of articles focuses solely on mainstream tourism journals. I therefore perused all mainstream tourism journals for which I could access the full text of articles in electronic format (this included *Annals of Tourism Research*, *Tourism Management*, *Journal of Travel Research*, *Tourism Culture and Communication*, *Current Issues in Tourism*, *Tourism Geographies*, *Tourist Studies*) using the keywords search 'visual methods', 'visual' and 'visuals'. The search included articles published from the 1970s through to 2010 but only articles from the 1990s could be accessed in full text format (prior to this only abstracts and not the full text could be obtained). Over 1,300 articles across the mentioned journals contained one or more of these key words but of those for which the full text could be accessed it was evident that the word visual was not being used in the context of visual research but in the context of the everyday usage of the term. For the overview or snapshot of articles used in this chapter, a total of thirty-three articles from the 1990s to 2010 across the mentioned journals were read in depth.

2 Indeed in my exegesis of the journal articles in tourism which utilised visual methods there was hardly any mention of the ethics involved in data collection for visual tourism projects. A notable exception however is the discussion of ethics contained in the article by Campelo *et al.* (2011) in which they discuss visual rhetoric and the ethics involved in the marketing of destinations. Specifically they argue for the 'necessity for ethical protocols and procedures to underpin the visual representations of identities, peoples and culture' (ibid.: 4).

Annotated further reading

Jacobsen, J. K. S. (2007) 'Use of landscape perception methods in tourism studies: a review of photo-based research approaches', *Tourism Geographies*, 9: 234–53.
This article is useful for those students and researchers who wish to apply insights from environmental psychology and photography in visual tourism research. It provides a useful review of landscape perception methods and also importantly contains a discussion of issues of validity within this context.

Markwell, K. (1997) 'Dimensions of photography in a nature based tour', *Annals of Tourism Research*, 24: 131–55.
This article is one of a few within visual tourism research which utilises a range of ethnographic techniques in order to explore the spatial, temporal and social dimensions of photography within the context of a nature based tour. It is also interesting in so far as it utilises content-analytic techniques in this context thus representing a combination of both quantitative and qualitative approaches.

Stanczak, G. (2007) V*isual Research Methods: Image, Society and Representation*. London: Sage.
While this text is not about the use of visuals within tourism, it addresses the question of epistemology within visual research studies in general. In this text, the interdisciplinary nature of visual research is highlighted. Students and researchers of tourism might be particularly interested in the introductory chapter by the editor as well as the chapters by Jon Wagner on documentary photography and by Barry Goldstein on why photos lie.

Yan, G. and Santos, C. A. (2009) '"China Forever": tourism discourse and self Orientalism', *Annals of Tourism Research*, 36: 295–315.
This journal article utilises video (which is under-utilised in visual tourism studies) to create new understandings about the way in which a tourism promotional video can serve as a mediated space in which representations of national identity are constructed. Students and researchers who are interested in postcolonial theory and the use of critical and discourse analysis within visual tourism research will find this article of interest.

References

Banks, M. (2001) *Visual Methods in Social Research*. London: Sage.

Burns, P. (2004) 'Six postcards from Arabia: a visual discourse of colonial travels in the Orient', *Tourist Studies*, 4: 255–75.

Campelo, A., Aitken, R., and Gnoth, J. (2011) 'Visual rhetoric and ethics in marketing of destinations', *Journal of Travel Research*, 50: 3–14.

Caton, K. and Santos, C. A. (2007) 'Closing the hermeneutic circle? Photographic encounters with the other', *Annals of Tourism Research*, 35: 7–26.

Crouch, D. and Lubben, N. (eds) (2003) *Visual Culture and Tourism*. Oxford: Berg.

Echtner, C. and Prasad, P. (2003) 'The context of Third World tourism marketing', *Annals of Tourism Research*, 30: 660–82.

Fairweather, J. R. and Swaffield, S. R. (2001) 'Visitor experiences of Kaikoura, New Zealand: an interpretive study using photographs of landscapes and Q method', *Tourism Management*, 22: 219–28.

Feighey, W. (2003) 'Negative image? Developing the visual in tourism research', *Current Issues in Tourism*, 6: 76–85.

Frost, W. (2010) 'Life changing experiences: film and tourists in the Australian outback', *Annals of Tourism Research*, 37: 707–26.

Garrod, B. (2008) 'Exploring place perception: a photo based analysis', *Annals of Tourism Research*, 35: 381–401.

Goldstein, B. (2007) 'All photos lie: images as data', in G. C. Stanczak (ed.), *Visual Research Methods*. London: Sage.

Hollinshead, K. (1999) 'Tourism as public culture: Horne's ideological commentary on the legerdemain of tourism', *International Journal of Tourism Research*, 1: 267–92.

Howarth, D. (2000) *Discourse*. Buckingham: Open University.

Humberstone, B. (1997) 'Challenging dominant ideologies in the research process', in C. Clarke and B. Humberstone (eds) *Researching Women and Sport*. London: MacMillan.

Jacobsen, J. K. S. (2007) 'Use of landscape perception methods in tourism studies: a review of photobased research approaches', *Tourism Geographies*, 9: 234–53.

Jamal, T. and Hollinshead, K. (2001) 'Tourism and the forbidden zone: the underserved power of qualitative inquiry', *Tourism Management*, 22: 63–82.

Kim. H. and Richardson, S. L. (2003) 'Motion picture impacts on destination images', *Annals of Tourism Research*, 30: 216–37.

Kirk, J. and Miller, M. L. (1986) *Reliability and Validity in Qualitative Research*. Beverly Hills: Sage Publications.

Lincoln, Y. S., and Guba, E. G. (1985) *Naturalistic Inquiry*. Beverly Hills, CA: Sage.

Lowenthal, D. (1985) *The Past is a Foreign Country*. Cambridge: Cambridge University Press.

MacKay, K. and Smith, M. C. (2006) 'Destination advertising: age and format effects on memory', *Annals of Tourism Research*, 33: 7–24.

Markwell, K. (1997) 'Dimensions of photography in a nature based tour', *Annals of Tourism Research*, 24: 131–55.

Markwick, M. (2001) 'Postcards from Malta: image, consumption, context', *Annals of Tourism Research*, 28: 417–38.

Morse, J. M., Barrett, M., Mayan, M., Olson, K., and Spiers, J. (2002) 'Verification strategies for establishing reliability and validity in qualitative research', *International Journal of Qualitative Methods*, 1: Retrieved 5 April 2010 from www.ualberta.ca/~ijqm.

Murphy, E., Dingwall, R., Greatbatch, D., Parker, S., and Watson, P. (1998) 'Qualitative research methods in health technology assessment: a review of the literature', *Health Technology Assessment*, 2 (16): 1–275.

Naoi, T., Airey, D., Iijima, S. and Niininen, O. (2006) 'Visitors' evaluation of an historical district: repertory grid analysis and laddering analysis with photographs', *Tourism Management*, 27: 420–36.

Norris Nicholson, H. (2006) 'Through the Balkan states: home movies as travel texts and tourism histories in the Mediterranean c. 1923–39', *Tourist Studies*, 6: 13–36.

Palmer, C. and Lester, J. A. (2007) 'Stalking the cannibals: photographic behaviour on the Sepik River', *Tourist Studies*, 7: 83–106.

Rojek, C. and Urry, J. (eds) (1997) *Touring Cultures: Transformations of Travel and Theory*. London: Routledge.

Said, E. (1995) *Orientalism: Western Conceptions of the Orient*. London: Penguin Books.

Schwandt, T. A. (1994) 'Constructivist, interpretivist approaches to human inquiry', in Denzin, N. and Lincoln, Y. (eds) *Handbook of Qualitative Research*. London: Sage.

Selwyn, T. (ed.) (1996) *The Tourist Image: Myths and Myth Making in Tourism*. New York: Wiley.

Sirakaya, E. and Sonmez, S. (2000) 'Gender images in state tourism brochures: an overlooked area in socially responsible tourism marketing', *Journal of Travel Research*, 38: 353–62.

Stanczak, G. C. (ed.) (2007) *Visual Research Methods: Image, Society and Representation*. London: Sage.

Tonkiss, F. (1998) 'Analysing discourse', in C. Seale, (ed.) *Researching Society and Culture*. London: Sage.

Tribe, J. (1997) 'The indiscipline of tourism', *Annals of Tourism Research*, 24: 638–57.

Tribe , J. (2008) 'The art of tourism', *Annals of Tourism Research*, 35: 924–44.

Tussyadiah, I. and Fesenmaier, D. (2009) 'Mediating tourist experiences: access to places via shared videos', *Annals of Tourism Research*, 36: 24–40.

Urry, J. (1990) *The Tourist Gaze*. London: Sage

Wagner, J. (2007) 'Observing culture and social life: documentary photography, fieldwork and social research', in G. C. Stanczak (ed.) *Visual Research Methods: Image, Society and Representation*. London: Sage.

Yan, G. and Santos, C. A. (2009). '"China Forever": tourism discourse and self orientalism', *Annals of Tourism Research*, 36: 295–315.

Part 3

Methods

4 Collecting visual materials from secondary sources

Salla Jokela and Pauliina Raento

Introduction

A growing portion of the information steering tourists' decision-making and behaviour is in visual form. Promotional images decorate brochures, websites, and postcards. These media offer valuable secondary sources for research purposes, because they shed light on the values and desires of tourism promoters and tourists. These sources are called "secondary," because those who created them are not involved in the research process.

In this chapter, we will discuss themes related to the collection of visual materials from secondary sources. The examples will illustrate how to choose and access secondary sources that fit a particular research purpose and how to justify the choice of material by contextualizing it in previous research and society. We will also address saturation and representativeness of data and ways of ordering the data collection process and the collected material.

Five examples

Secondary source materials can be collected from multiple media at various times and places. The goal is to reach a manageable, reliable, and representative sample that fits the purpose in question. But where to start and how much material is needed? The agony is that there are no simple answers that would serve all projects. Instead, new questions emerge as the project proceeds. They need careful attention, for a solid process of data collection is a necessary prerequisite for success in the rest of the research project, discussed in the other chapters of this book.

We will exemplify the collection of visual materials from secondary sources with five types of data, which are accessible and inexpensive because they are mass-produced: (1) destination brochures; (2) postcards; (3) postage stamps; (4) the Internet; and (5) historical landscape photographs. These are a staple part of tourists' daily routines, and therefore global in character. Yet these data tell unique stories, for each place, country, or company within the universal framework of tourism and destination marketing has its own "personality." Connections to multiple scales and processes relevant to society and the tourism industry make these source materials suitable for several purposes. The amount of material can be adjusted to serve a show-and-tell, a term paper, a thesis, and a full scholarly research project with ambitious publishing goals. These five types of data are student- and teacher-friendly, but their collection demands rigour to make the abundance manageable.

Another characteristic that these secondary source materials share is a connection to broad soci[et]al issues, ideologies, and to the ways these work. In the case of commonly used visual materials, relevant contexts include the promotion, marketing, and image-making of places; representation of destinations and lifestyles, national or other ideologies and shared values; and the basic mechanisms of propaganda and persuasion, which include both political messages and marketing talk. These frameworks have been applied previously in interdisciplinary tourism studies to the same or similar materials. Photographs have been studied extensively (e.g. Osborne 2000; Jenkins 2003), and some research exists on destination brochures and postcards (e.g. Edwards 1996; Buzinde *et al.* 2006). The Internet is attracting attention, as tourism advertisement and booking of trips have moved online (e.g. Hashim *et al.* 2007; Kim and Fesenmaier 2008). In turn, very little has been said in tourism research about postage stamps, even if their pictures have depicted tourism-related themes for over 100 years (Raento 2009a). Knowing what kind of studies exist about similar materials is necessary, for previous research always supports new projects. What others have written helps in defining the foci of a new project, by steering research questions or solutions to particular dilemmas. Each data set is nevertheless unique so that a scholar must answer specific questions and make defendable, informed choices about that material. The material itself often helps in developing the project so that the process moves simultaneously in multiple directions. All projects are learning processes, but good preparation can eliminate unnecessary errors.

We will ask questions that are typical of the data collection process. We use our own research and therefore Finland as an example, but what we say is broadly applicable to other types of materials, sources, and places. Our work—and these guidelines—is supported by a research philosophy that can be summarized in four points. First, we believe that what people do in their daily life matters for serious scholarly work. We therefore swear by the value of everyday "stuff" in teasing out the best potential of case studies and producing new, abstract knowledge about human societies. Second, we believe that research is, and should be, fun. Everyday "stuff" tends to meet this criterion because it often leads to brain-stimulating revelations ("Oh, I never thought about that, even if it's been under my nose all this time!"). Third, we believe that new ideas should be treated with intellectual curiosity and be given the opportunity to show their worth. We therefore deem important the overcoming of any remaining scholarly resistance against exploring this mundane "stuff." And fourth, we believe in the value of interdisciplinarity. Curiosity that ignores disciplinary boundaries often rewards by broadening insights and saves efforts by combining them (Raento 2009b).

We now turn to the five examples that illustrate the collection of visual materials from secondary sources. Tourism brochure images show how solid contextualization justifies the collection and definition of data. Postcards instruct how to create a representative sample from a seemingly infinite range of options. Postage stamps exemplify one way to collect a comprehensive sample that covers everything there is. Online sources highlight source criticism and suggest that the most convenient source may not be the most suitable one. Finally, landscape photographs address archival work, including issues about access, recording of information, and the interplay between research questions and data collection.

Images in tourism brochures

Tourists like brochures, for they are inexpensive (free), accessible, and easy to carry around. Their images inform about a destination at one glance. They also reinforce place-bound stereotypes and identities and tell about the values and ideologies of their producers and consumers. Pictures in brochures shape the ways in which tourists behave in, and look at, a destination's landscape. Brochures may also function as souvenirs that distribute ideas and images of a destination to the tourist's home community. To understand these mechanisms students of tourism brochures must ask, already at the data collection stage, *who* produced their images, *why* this was done, and *in what context* all this happened.

Practical Tips 4.1

GET THE BIG PICTURE

Comprehensive groundwork that supports the entire study process starts at the early stages of data collection. The following three "commands" have made us perform better:

- **Read!** Reading broadly helps in understanding the data. Start by identifying the most relevant sources in order to manage the flood of information. Think creatively about keywords and learn to use library and research databases to discover what is out there. Many books can be previewed online.
- **Look around!** Good knowledge about cultural visual references – popular or otherwise – aids in comprehending the multiple messages embedded in images and in creating solid data sets. Learn about symbols, motifs, and the meanings of stylistic devices.
- **Listen!** Valuable background information about the data can be available in oral form. Many producers of visual materials are willing to discuss their goals and world views. Ask politely for some time to chat and you may be rewarded with new insights.

A study about the tourism brochure images of Finland's capital city Helsinki shows how data collection is tied to data analysis and how preparation is needed before collecting can begin. The project started from a general interest in tourism, political landscape studies, visual methodologies, and the history of Finland. Studies about these topics elsewhere showed that tourism promotion was frequently used as an identity-political tool during historical transition periods (e.g. Morgan and Pritchard 1998; Ateljevic and Doorne 2002). Further contextualization—inquiries into Finland's complex political history between East (Russia) and West (Sweden)—raised a question of how tourism promotion was used in this country and what would explain its particular applications. The focus settled on Helsinki, because its tourism imagery was little studied despite the capital city's historical role as the "façade for Finland" (Jokela 2011). What kind of source materials would best illustrate these processes in this case?

An overview of promotional publications of Helsinki was available at the National Library of Finland, which houses a large collection of tourism brochures, itineraries, guidebooks, and catalogues. This abundance of data encouraged the limitation of the study to a manageable, yet analytically interesting sample of brochures. The data selection process concluded with ten brochures (with a total of 127 images), published by the Excursion Section of the Sport and Excursion Office of the City of Helsinki. The selection was data-driven and relied on grounding the project in previous research. Consultation of previous research thus supported sound choices in data collection.

A scholar should be able to defend a particular data collection process and to explain the choices so that the exercise can be evaluated. This imagery of ten brochures was deemed suitable for the investigation of tourism promotion in changing political and cultural circumstances for four key reasons. The brochures were published in the 1950s and 1960s, when Finland was recovering from World War II. In the context of the Cold War, Finland adjusted to a new geopolitical situation and established close relations with the Soviet Union, the former enemy. A need to nurture Finnish national identity and to create a distinct, internationally attractive image for Helsinki followed the political change. The globally booming tourism industry facilitated the promotion of Helsinki as a thriving destination especially for Western tourists.

Second, the Sport and Excursion Office was the leading producer of Helsinki tourism brochures after the war. Third, publications of the City of Helsinki offered valuable contextualizing information about the operation of the Office and the circulation and language selection of the brochures. By way of example, these supporting materials revealed that the Office was the first agency designated to specifically attend to the needs of tourists. This confirmed that the material indeed represented the official tourism promotion of Helsinki. Statistics about the edition size, and the multiple languages used in the brochures, led to the assumption that these images represented what tourists encountered when they searched for information about Helsinki.

Fourth, these brochures comprised the first coherent and extensively illustrated brochure series of the city.

The example shows how early and comprehensive contextualization of data in research and society supports its collection, because this helps to decide what should be included in, and what excluded from, the study. Once the choice of material is justified, a representative sample can be collected efficiently and defended at all stages of the study.

Postcards

Postcards are a staple part of global tourism. Tourists send them to family and friends despite the possibilities of electronic communication. Tourism promoters and place marketers in every destination therefore produce postcards that illustrate the place and cater to a variety of tastes. The imagery on postcards reflects the techniques its producers use in place-promotion—and the preferences and perceptions of their consumers. The often massive number of choices challenges a student of postcards to collect a representative sample. Going beyond a random or hastily collected "convenience sample" requires thinking about, and making, defendable choices about the data collection process. One goal is to manage time, money, and effort.

Three basic questions—*when*, *where*, and *how*—gave order to the data collection process in an inquiry into the representation and marketing of Helsinki in postcard imagery. As an answer to the first question, the collection was scheduled to take place at the height of the local tourism season and before a major international tourism event in central Helsinki—the World Championships in Track and Field Athletics at the Olympic Stadium in the summer of 2005. The reasoning went that at that time the broadest possible selection of postcards would be available for a tourist touring the city. Statistics about typical visitation length to Helsinki defended the collection period of a few days (which also made things convenient for the collector)—this was the expected time a foreign tourist would spend in town. The selection would be at its widest because the postcard dealers would stock up in anticipation of peak sales during a season when visitation was generally high. Informal chats with the postcard vendors during data collection confirmed that this assumption was correct.

The question *where* pointed to kiosks, souvenir shops, and book stores. What locations would best support the collection of a representative sample? The process started with thinking about where tourists go to buy postcards. Statistics about visitation to individual sites and first-hand long-term observations about what tourists do in Helsinki helped in choosing locations. Chats with friends about where one would guide tourists to buy postcards in central Helsinki confirmed the choices. Types of tourists and travel budgets were factored in loosely so that there was variation in the pricing of postcards and style of vendors. Based on these factors six initial collection points were selected: a historic market hall in a neighbourhood with bohemian-professional, student-alternative, and recent immigrant flavours; the main post office and an upscale bookstore in the core of the city; a souvenir store complex next to several popular sites off the immediate Downtown area; a UNESCO world heritage outdoors destination, with multiple little shops and kiosks of various characters; and a busy kiosk at the central railway station.

The guiding principles of *how* were representation and saturation. The goal was to create an exhaustive sample of those Helsinki-depicting postcards that were available at the moment of visitation in these locations. This sample would represent the maximum selection available to the tourist during an average-length visit. The available images would only be counted in (and purchased) once, but notes were taken about repeated appearance. The idea of saturation meant that the sample could be considered representative when new locations no longer significantly added to the sample. Therefore adding locations was an option, if the six places would not suffice. But this did not happen, for the last two collection points had very few new images. At this point of saturation the collection included some 300 postcards, which was a reasonable number for analysis and within the budget. It helped that the managers of those stores where the largest purchases were made were willing to negotiate a bulk price in support of scholarly work. During the time-consuming scanning of the available imagery, the vendors discussed the production and distribution of tourism postcards in Helsinki. This was valuable background information at the stage of data analysis.

Postage stamps

The sending of a postcard requires a postage stamp, which is one overlooked but powerful data source in visual tourism studies. Stamps are official documents of the issuing state and therefore their images reflect the sources of national pride in that country. Because stamps communicate to international audiences via global mail, their pictures can promote places and impressions. What makes stamps particularly useful for examining national representations and promotional campaigns is that they form spatially and temporally comprehensive, easily accessible, and comparable samples. In other words, everything that there is, and has ever been, can be included and then split into smaller data sets for specific purposes. The typical number of stamps per country ranges from one to three thousand. From these wholes smaller comparable samples can be created by limiting time or theme.

So how does one gather a complete set of stamps without going bankrupt at the collector store? A search on the Internet, and a chat at the post office and at a collector's store, will confirm that information about a country's postage stamps is easy to find. National postal services have stamp-issuing programs, under which dozens of new stamps are issued every year. Selling them to collectors is good business for the issuing country and private entrepreneurs, so both postal services and major philatelic dealers publish yearbooks and catalogues about what is available. They typically include a colour image and information of each and every stamp issued in a country or region in a given time. For scholarly purposes an official (or widely used and trusted commercial) catalogue suffices, especially if its contents are checked against the collection of originals at the postal museum. Additional information can be found through the websites and publications of national postal services (e.g. www.posti.fi and the free *Stamp Info* in Finland) and of collectors (e.g. www.filatelia.net and *Suomen Postimerkkilehti*, a magazine by Finnish philatelists). A selection of original stamps can be purchased for a closer look and illustration purposes once the study proceeds. In terms of access to complete sets of source material, stamps resemble travel magazines, the full volumes of which are easy to gather (see Lutz and Collins 1993, about a study of photographs in *National Geographic*).

A study about tourism and nationalism on Finnish stamps relied on the above-described sources (Raento 2009a). The data was collected from the annual commercial Finnish stamp catalogue LAPE, but cross-checked against the originals at the Postal Museum in Helsinki. The collection process zoomed in toward the final sample at two phases. The first phase formed part of a broader study about visual methodologies and political geography, for which a systematic sample of Finnish stamps was collected (see Raento and Brunn 2005). The first task was to determine where to start and end the collection, for Finland's stamp history dates back to 1856 and new stamps are issued every year. Because of the complex political history of Finnish stamps, a decision was made to ignore all stamps issued before independence and to start from the first markka-valued issue of independent Finland in 1917. The change of currency from the national markka to the euro in the European Union in 2002 provided a convenient end year. The focus on state-wide public access helped in justifying the exclusion of all those stamps that had been issued for special purposes (e.g. type of cargo) or regions (e.g. autonomy, occupied territory). Because the focus was on the images, philatelic details were ignored (e.g. multiple issues with the same picture but different watermarks or perforation). The first phase of data collection thus included all markka-valued stamps of independent continental Finland that depicted different images: 1,501 stamps from 1917 to the end of 2001.

In the second phase the sample was narrowed down to study images about tourism, leisure, and recreation. The stamps were selected by determining the primary theme in each image. Because many images carry multiple messages, this challenge required the cross-checking of multiple supporting sources, including official information about the purpose of each issue. The guiding principle was that each decision to include or exclude an image would have to be defendable and in line with the other decisions about the selection. One tough decision was to exclude all bird stamps despite the popularity of bird-watching and the known value of these stamps as bird-watchers' souvenirs and collectables. The argument was that these stamps were primarily zoological and these birds were not exclusive to Finland (even if typical of its fauna), so that one would not have to travel to Finland to see them.

Figure 4.1 A Finnish postage stamp (issued in 1947)

Some decisions about what to include and what to exclude from a set of visual material are more obvious than others. A Finnish postage stamp, issued in 1947, exemplifies why broad background knowledge about the society surrounding the data is necessary in the collection of visual materials from secondary sources (see Figure 4.1). Knowing that Koli, the place named on the stamp, is a nationally important landscape and tourism destination in Finland, connects the lake view to the tourism industry and to Finnish nation-building. The Finnish- and Swedish-language text reveals that the stamp celebrates an anniversary of the Tourist Society in Finland. Text often accompanies images and adds to their message. Sometimes access to a foreign language may be needed to get the full picture. This image of a natural landscape was included in a study about Finnish tourism, nation-building, and postage stamps because of its multiple connections to tourism (Raento 2009a).

Through this reasoning, the data gradually fell under seven categories that simultaneously supported the ordering of the data collection. They covered times of leisure and recreation (e.g. holidays); sites in cities and rural areas; wilderness; and various activities related to outdoors and relaxation. Explicit promotions of tourism were also included, and so was philately (including international stamp exhibitions). The second phase resulted in a total of 381 stamps, the data for the study about tourism and nationalism on Finnish postage stamps (Raento 2009a).

Online materials

The Internet has shaken up the world of tourism by facilitating immediate, easy distribution of information across the globe and by creating virtual leisure options and communities. Many "traditional" source materials suitable for visual analysis are now online: tourism brochures, advertisements, electronic postcards, and illustrated travel blogs are among the examples.

One important challenge in the collection of online materials is to decide how to choose from a zillion hits on Google. Here geographical scale can help. In a project that examined how Finland is being promoted online, the data had to concern the entire country (national scale) and be produced by national tourism authorities (official image). Behind this focus was our interest in the role of tourism in state-led place-promotion and image-making. Following this reasoning the *Visit Finland* Internet site, by the Finnish Tourist Board, was chosen for closer inspection, as it is the most important site for "official" state-led tourism promotion in Finland (www.visitfinland.com).

That the Internet is a "democratic" forum where anyone can have a voice can both help and burden an online data collector. On the one hand, online source materials offer access to the study of those meanings that users attach to tourism destinations and their visual representations. For example, users of *Visit Finland* can share their experiences online and download photos of Finland. On the other hand, everything might not be what it first seems, and verifying who is behind a particular site can be difficult. The source can be biased and abundance can complicate reliable comparisons. Therefore determining what (or whom) the material represents can be difficult. Site administrators choose what is published on a site and emphasised in its layout. Many users of tourism-related sites are affluent individuals, who keep pace with technological change and can afford to travel, and therefore represent only a fraction of all people touched by this medium. Furthermore, Internet access is uneven, as some countries impose restrictions on their citizens' online activities or censor certain sites. These issues of bias and representativeness are reasons why online materials underscore the need for source criticism already at the data collection stage.

Caution is needed also because of the immediate nature of the Internet, where things seem to happen *now* in one big global space. Things constantly change online. This "instability" poses a data storage challenge to the collector of online source materials. Web page contents change or sites disappear altogether, which complicates finding the same materials even shortly after their initial collection. This means that the data collector must preserve all data in the version that was available at the moment of collection. Internet sites can be stored in multiple ways. One way is to print the pages, which may ease their comparison but is wasteful—and many sites contain multimedia and flash shows that do not print well. In our examination of the *Visit Finland* site, these visualizations were stored by taking snapshots of the screen (using the Print Screen key). The pictures were then pasted onto a photo-editing program and saved for further use. This guaranteed that the entire data set stayed unchanged and represented "the big picture" in a given, concentrated moment of data collection.

Figure 4.2 Screen snapshot

Source: www.visitfinland.com (accessed December 2009), used with permission

Taking snapshots of the screen is a practical way to record Internet pages without losing visual information (see Figure 4.2). Snapshots reproduce the organization of visual elements in the original material. This is one way to meet the scholar's responsibility of storing the data in its original form.

The short history of the Internet, together with its changing nature, highlights the present time. This means that studies based on online sources typically examine what happens "now" or otherwise cover a very limited time period, even though some old web pages can be accessed through an online archive (www.archive.org/index.php). This means that the Internet may not be suitable as a data source for projects that involve a historical perspective or require an in-depth understanding of long-term change. Even if the Internet is accessible

Practical Tips 4.2

WHAT IS REPRESENTATIVE ANYWAY?

What makes data representative, and how much data is enough, are difficult questions. At one extreme the postage stamps can cover everything there is with a reasonable amount of work. At the other extreme, the Internet offers infinite options.

How much data is needed for credible representativeness depends on multiple factors. These may include access issues, purpose of study, and research questions. An in-depth content analysis of tourism-promotional imagery for methodological purposes is a different project from an empirical study of a destination's image. Answers are always case-specific.

The following questions aid in determining the limits and limitations of a data set:

- What is the data supposed to represent?
- What can this particular data be used for?
- What answers can this data give?
- What kind of conclusions can this data support?

These questions must be asked (and often answered) already at the data collection stage, because they steer both methods and results.

at home or in the nearest library, and is very convenient to use, it may not be the right data source for all projects. Matching a secondary source to a project is therefore a crucial part of planning the data collection process.

As a relatively new medium the Internet also highlights changes in ways scholars think about source materials and their collection. Generation and technological savvy are important factors in this regard. For today's teenagers the logic of the Internet is often readily more legible than to their grandparents. In any case, one challenge that needs to be considered in the gathering of online sources is to figure out how a website is structured and how a typical Internet user navigates on it. The procedure is typically non-linear so that the users concentrate on searching for specific information or click spontaneously on icons and links that please them. This fragmentation highlights the importance of well-formulated research questions that support the data collection: What is it that one is supposed to find out, and look for, in this particular project?

Our aim was to understand in a general way the use, functions, and mutual relationships of visual elements on a national country-promoting website. We first proceeded to find out that *Visit Finland* has a multi-centred and hierarchical order, like most Internet sites, which greatly helped in comprehending the whole. A site map linked to the front page explained the structure. The main titles indicated what themes the tourism promoters responsible for the site considered to be central, while secondary issues were listed underneath. In order to harvest all relevant information from the site in a given moment, we saved the content of all these pages in a systematic way, making sure that our storage system preserved the individual pages' mutual order, hierarchies, and layout.

Landscape photographs

Gatekeepers sometimes control access to secondary sources of visual materials. These people between scholars and the data include archivists whose job is to preserve rare and valuable materials. These people decide the conditions for using an archival collection, and control documentation and information flows. To guarantee successful collaboration the scholar wanting to use an archive must obey its rules and behave in a respectful manner. Good planning is of paramount importance, for one may have to explain in plain language why access to particular source materials is needed or what kind of material one hopes to find in the collection.

Archival work was necessary to carry out a study about the role of historical landscape photographs in Finnish tourism promotion (Jokela and Linkola 2009). The project explored how photos were used to enhance national solidarity and consciousness in a newly independent country which, before 1917, had belonged to Imperial Russia. After an inspection of virtual and manual databases supplied by various archives, a closer look was taken at the collections of the National Archives of Finland. These housed an extensive set of printed papers of the Tourist Society in Finland. Research literature about the history of tourism in Finland had confirmed that this society was the most important national tourism promoter during the examined time period, so this source appeared to be the most logical choice. Of particular interest was a series of eleven guidebooks from the 1920s, which comprised the first systematic and extensive body of illustrated promotional publications about Finland.

Sorting through the archive was easy, because the folders containing the source materials were brought to a reading room at a few hours advance notice and they could be handled without mandatory supervision of an archivist. However, patience and creative thinking were needed to figure out the ordering logic of the old collections and to gather the desired material. It turned out that the material was catalogued under the year in which the archive had received it, rather than per year of publication or per theme. Accordingly, the guidebook material was split between several somewhat randomly ordered folders and buried under other items that had no relevance to this project. The rules of the archive obliged the visitors to keep all materials in their original order, which meant that there was no fast way out (such as rearranging the contents of some folders).

Practical Tips 4.3

GOOD RECORD-KEEPING PAYS OFF

Good record-keeping at data collection stage saves time, money, and nerves. The following guidelines have helped in our projects:

- Creative adjustment to circumstances is cost-efficient. A pencil holds better on paper than a pen in cold climes. A (cell phone) camera can be faster and environmentally sounder than the copy machine.
- What now seems obvious may be hard to figure out later. File data systematically and go back patiently if you change or forget something. Notes about how the material was organized will help finding it later.
- It is easier to delete unnecessary notes than to keep going back to the source to double-check things. Pay attention to detail and record them. Match individual data items, locations, and times of collection.
- Leaving records hanging guarantees a loss of valuable information. Write out and organize all notes immediately after each assignment.

The first task, therefore, was to record in detail the contents of each folder to facilitate the data collection. Permission to use a digital camera and a laptop for record-keeping were obtained from the staff. This technology had several advantages in book-keeping. One was its speed, another was the low cost of photographing compared to photocopying. It also felt like an ethically sound choice, for both the camera and the laptop saved plenty of paper (that would eventually turn into waste). Avoiding the copy machine also protected the delicate old documents from bending, which made the archivist happy. In order to avoid the trouble of returning to the archive to patch up the data set, every item was recorded at this stage of data collection. It is typical of data processing that certain details seem insignificant at first, but turn out to be valuable later. This thoroughness is even more important if the archive is far away or in a cultural setting where the understanding of meaningful details may require additional background work.

In the end, the laborious logic of the archival collection turned out to be a blessing in disguise. Browsing through various publications offered information that was necessary in the data collection process. For example, the Yearbooks of the Tourist Society in Finland contained details about the production of the examined guidebook series. These publications confirmed that all relevant materials had indeed been collected. Very importantly, unexpected encounters helped in developing and focusing the original research question about the use of touristic photos in civic education for national unity. For example, a 1920 issue of the Finnish-language *Tourist Magazine* (*Matkailulehti*) included a contribution by a prominent Finnish geographer

Ragnar Numelin, who stated that "tourism should work for geography's favor, or vice versa." Implying a close and early link between the goals of tourism promotion and academic geography in Finland, this finding demanded a more-detailed-than-anticipated look at the connections between the landscape imageries used in Finnish tourism guidebooks and popular-geographical publications in the 1920s.

The example shows why it is important to accept that an archive's logic may not match that of the scholar, but valuable new knowledge may reward patience and curiosity. Small clues that reflect the spirit of an age and past ways of thinking may add value to the collected material or complement it. These past ways may have been self-evident for the contemporaries who produced the scholar's source materials, but a scholar who lives in the present time may have to work hard to understand them. This "big picture" typically comes together piece by piece. Intuition, empathy, and imagination are therefore valuable assets in systematic data collection from secondary sources.

Ethical issues

No data exists in a vacuum, for someone produces it. In the case of secondary data sources, choices of themes, layout, and media are embedded in the data in multiple layers—from the production decisions of these materials to the scholar who collects, examines, and interprets them. These processes take place in particular times, spaces, and societies, to which they need to be connected and against which they must be interpreted. The requirement to think critically about connections also applies to the scholars themselves, because choices are always subjective. The cultural, political, and socio-economic background and past experiences of individuals condition their decisions and ways to see the world. This possibility of bias—or the ability to see (or not see) something—must be addressed in data collection.

The reliability of the process can be strengthened by performing what tourism researchers Echtner and Prasad (2003) have called "cross-checks." In these check-ups different sources of information are placed in dialogue with one another to cross-examine each other's credibility. This is particularly necessary in the case of visual data research, where under examination are representations of culture, gender, and ethnicity; perceptions and prejudices of hosts and visitors; and power relations and their consequences. One way to control the inherent subjectivity of one's mind is to collaborate with other researchers, perhaps across cultural, national, and disciplinary boundaries, and openly discuss the possibility of bias. In a study about postage stamps a comparative critical discussion between two scholars opened new insights at every stage of the research process (Raento and Brunn 2005). The participants came from the same academic field and professional backgrounds of similar international experience, but were of different nationality, gender, and generation. An explicit acknowledgement and thinking through of the impact of differently conditioned perspectives greatly increased the credibility of the project and its results.

Another way to perform "cross-checks" is to employ a broad range of background materials in support of the data collection process. The discussed examples show how first-hand knowledge of the study environment was combined with information gathered through discussion, statistics, and online and other searches. Previous scholarly work helped in making decisions about data collection by offering possible, tested solutions to particular problems. A successful and responsible collection process thus relies on the power of *intertextuality*—on the explicit, critical discourse between multiple sources, which includes the researcher's understanding of his/her subjective self.

One's behaviour is another ethical issue. It requires finding out about, and obeying, rules in archives and similarly respectful conduct in other data collection sites, be they corner kiosks, souvenir stores, or administrative offices. Considerate behaviour includes civil and culturally appropriate treatment of people, and the making of fair deals with vendors. Success of the collection of secondary source materials often depends on common sense and sensitivity to changing situations. Friendliness, politeness, and cultural caution are readily accessible assets of all scholars—whose responsibility it is to employ them.

Appropriate conduct requires attention from the first steps of data collection, for access to data is rarely free from ethical considerations. Tough questions follow: how many free brochures can one just go and grab? Should one, in a given situation, reveal scholarly interests or pretend to be an information-hungry tourist? What, if any, is the impact of these choices on the data set and expected results of the study? How to access private, personal, or unpublished materials behind permissions, recommendations, or transfers of money? What does one do with an organizational culture that is suspicious of nosy outsiders or prone to avoid the (perceived) extra work? What can one comfortably accept and adjust to, and under what circumstances, to gain access to particular data? Should one focus on building social trust or using hard currency (or neither)—and where is the line of exploitation? How valuable or irreplaceable is a particular set of data? Why so? What are the costs and benefits of a chosen route for each involved party?

Situations and answers vary, but knowledge about cultural codes and social sensitivity support ethically sustainable choices that hold under critical scrutiny. Talking with colleagues about their experiences may be of great help.

CHAPTER SUMMARY

Our key message about collecting visual materials from secondary sources is summarized by the following points:

- Choices in data collection are inherently subjective and must be defensible. Smooth progress requires careful planning and critical thinking. Rigour, patience, and source criticism will be rewarded.

- Data collection is connected to previous scholarly tradition and society. Earlier studies help in clarifying research questions and data collection choices. These often evolve as the collection project proceeds.
- Saturation, representativeness, and comprehensiveness are common dilemmas of each data collection process. Solutions are often data and case specific and demand creative thinking, for each set of visual research material is unique.
- Many visual materials are free of charge and openly accessible, but the most convenient choice may not be the best one. Negotiating access—like all steps of data collection—requires thinking about ethical issues and one's own research philosophy.

Annotated further reading

Bhattacharyya, D. (1997) "Mediating India: an analysis of a guidebook," *Annals of Tourism Research*, 24: 371–389.
This article highlights the value of looking at both visual images and text, from the perspective of their mutually enhancing interaction. The case study illustrates how to examine destination guide books for scholarly purposes.

Fürsich, E. (2002) "Packaging culture: the potential and limitations of travel programs on global television," *Communication Quarterly*, 50: 204–226.
Three globally distributed travel program series on the US-based Travel Channel serve as research data for this insightful study about the political and cultural dimensions of visual communication and entertainment.

Hanna, S. P. and Del Casino Jr., V. J. (eds) (2003) *Mapping Tourism*. Minneapolis: University of Minnesota Press.
The complex relationship between tourist maps, spaces, and identities is examined in the chapters of this edited volume. The contributions show the value of cartography as data, as a research tool, and as a result.

Lutz, C. A. and Collins, J. L. (1993) *Reading National Geographic*. Chicago: University of Chicago Press.
This book, by an anthropologist and a sociologist, addresses popular representations of the so-called Third World cultures. The critical study exemplifies how careful definition and sampling of visual data, and constant sensitivity to research objectives, aid in collecting a valid sample.

Rose, G. (2007) *Visual Methodologies: An Introduction to the Interpretation of Visual Materials. 2nd edition*. London: Sage.
This interdisciplinary and very popular book has its roots in human geography. The author outlines clearly and concisely the principles of data collection specific to various visual research methods. These range from semiology to discourse analysis.

References

Ateljevic, I. and Doorne, S. (2002) "Representing New Zealand: tourism imagery and ideology," *Annals of Tourism Research*, 29: 648–667.

Buzinde, C. N., Santos, C. A. and Smith, S. L. J. (2006) "Ethnic representations: destination imagery," *Annals of Tourism Research*, 33: 707–728.

Echtner, C. and Prasad, P. (2003) "The context of Third World tourism marketing," *Annals of Tourism Research*, 30: 660–682.

Edwards, E. (1996) "Postcards: greetings from another world," in T. Selwyn (ed.) *The Tourist Image: Myths and Myth Making in Tourism*, Chichester: John Wiley & Sons.

Hashim, N. H., Murphy, J. and Hashim N. M. (2007) "Islam and online imagery on Malaysian tourist destination websites," *Journal of Computer-Mediated Communication*, 12: 1082–1102.

Jenkins, O. (2003) "Photography and travel brochures: the circle of representation," *Tourism Geographies*, 5: 305–328.

Jokela, S. (2011) "Building a façade for Finland: Helsinki in tourism imagery," *The Geographical Review*, 101: 53–70.

Jokela, S. and Linkola, H. (2009) "Patriotic landscapes of tourism and geography: Finland in the 1920s," poster presented at the Association of American Geographers Annual Meeting, Las Vegas, NV (USA), March.

Kim, H. and Fesenmaier, D. R. (2008) "Persuasive design of destination web sites: an analysis of first impression," *Journal of Travel Research*, 47: 3–13.

Lutz, C. A. and Collins J. L. (1993) *Reading National Geographic*, Chicago: The University of Chicago Press.

Morgan, N. and Pritchard, A. (1998) *Tourism Promotion and Power: Creating Images, Creating Identities*. Chichester: John Wiley & Sons.

Numelin, R. (1920) "Maantiede ja matkailu" ("Geography and tourism"), *Matkailulehti* ("*Tourism Magazine*"), 1: 13–16.

Osborne, P. D. (2000) *Travelling Light: Photography, Travel and Visual Culture*. Manchester: Manchester University Press.

Raento, P. (2009a) "Tourism, nation, and the postage stamp: examples from Finland," *Annals of Tourism Research*, 36: 124–148.

—— (2009b) "Interdisciplinarity," in R. Kitchin and N. Thrift (eds.) *The International Encyclopedia of Human Geography, Volume "General"* (ed. J. Sidaway). Oxford: Elsevier.

Raento, P. and Brunn, S. D. (2005) "Visualizing Finland: postage stamps as political messengers," *Geografiska Annaler* B, 87: 145–163.

5 Eliciting embodied knowledge and response

Respondent-led photography and visual autoethnography

Caroline Scarles

Introduction

This chapter moves from notions of visuals as a means of introducing secondary data to attend to the roles of photographs as producing data. Drawing upon methods of respondent-led photo-elicitation and visual auto-ethnography, it resituates respondents as the producers, creators and indeed, directors of the visuals encountered during the research process. As Wang and Burris (1992) and Warren (2005) suggest the processes of producing and introducing photographs into the research setting gives respondents 'photo-voice' as power and control is renegotiated from researchers to respondents. This chapter explores the ways in which visuals (in particular photographs) become central to accessing embodied spaces of encounter as they not only offer respondents comfort and reassurance, but facilitate the 'connection' between researcher and respondent as knowledges are transferred and shared. Thus, visuals create spaces of understanding as the potential arises to transcend the limitations of verbal discourse and open spaces for creativity, reflection and comprehension. However, while realising the opportunities such methods afford, the chapter will also consider the limitations that inevitably arise through the use of such techniques.

From ocularcentrism to embodied visualities in tourism research

Photography and the visual have been fundamental to research on tourism since early studies into the ocularcentric practices of The Grand Tour (Löfgren 1999). Such practices positioned tourists and their visual techniques as, dis-engaged, detached beings who experienced places, cultures and people through overarching gazes and practices of observation. Creating a visual-isation of the travel experience (Adler 1989; Craik 1997; Urry 1990), intense ocularcentrism pervaded tourist practice and tourists were elevated as all-seeing authorities; colonising others through visual practice. Vision and the visual were secured as the key sense of the tourist encounter as tourists captured and recorded, controlled and categorised destinations. Such primacy continued to permeate understanding as Urry (1990) conceptualised the 'tourist gaze' where tourist spaces become understood through the practice of 'gazing' as objects are rendered worthy of attention through signposting, signification and meaning interpretation and are fixed both spatially and temporally.

Such positivist understanding of tourist behaviour and practice is paralleled in well-rehearsed visual methods such as content analysis (e.g., Dann 1988; Dilley 1986; Edwards 1996; Thurot and Thurot 1983) and semiotic analysis (e.g., Markwick 2001) of tourism media. However, recent years have witnessed dramatic shifts in theoretical understandings of tourism and what it means to be a tourist. Moving beyond understandings of tourism as dichotomies of work/play and home/abroad, authors such as Franklin and Crang (2001) propose that tourists no longer exist in spatially and temporally fixed locations bound by notions of seeking the 'authentic' other (see MacCannell 1973; Graburn 1977; Cohen 1988). They should no longer be thought of as leaving behind their everyday self; moving body, self and being to another location while adopting a different form of being and situatedness that is appropriate to their new, alien and somewhat rather exotic location. Rather, tourism becomes imbued in a web of complex performative processes and practices (see Rojek and Urry 1997; Crang 1997, 1999; Coleman and Crang 2002; Crouch 2000a, 2000b; Edensor 1998, 2000, 2001; and Franklin 2003). Tourism is a fluid and mobile process of becoming rather than a static state of being as tourists move through a series of spaces and continually reconfigure their selves as they encounter different places and cultures (see also Scarles 2009). It is a series of wholly multisensual encounters that embraces a plenitude of potential subjectivities and experiences and accesses the lay and popular knowledges of the tourist experience (Crouch 2000a, 2000b; Crang 1999).

As attention turns to the embodied performances of touristic encounters, authors such as Bennett (2004), Crang (1997, 2002, 2003) and Rose and Gregson (2000) call for innovative methodological approaches to address the emotional, sensual, embodied and performative nature of social practice. As Bijoux and Myers (2006: 46) suggest 'we cannot have a complete discussion of the experiences of bodies in place without considering the role of feeling, thoughts and emotion'. Therefore, while the value of methods such as content and semiotic analysis can never be denied, as authors such as Tribe (2004) call for greater intellectual space for 'new' research, we must be methodologically equipped to embark on such a journey. Indeed, as Scarles (2010: 2) suggests, 'alternative methods are required that engage with research participants in ways that move beyond the realms of representation to access the haptic, non-representational spaces of encounter and experience'.

It is therefore important to address the role of the visual as a tool for accessing and mobilising affectual and embodied expressions of self. As Pink (2007: 21) suggests:

> visual research methods are not purely visual. Rather, they pay particular attention to visual aspects of culture. Similarly, they cannot be used

> independently of other methods; neither a purely visual ethnography nor an exclusively visual approach to culture can exist.

It is important to realise therefore that the visual is more-than-can-be-seen. Moving beyond paradigms of ocularcentrism (Jay 1997), the visual emerges as integral to other sensual modalities. As Bærenholdt *et al.* (2004), Crouch (2000a, 2000b) and Veijola and Jokinen (1994) suggest, in researching tourist practice and performance, we must embrace the plurality of sensual interplays of tourist practice as subjective, reflexive and poetic occurrences and utterances of self and other.

Thus, photographs, as visuals, are no longer 'static, distanced and disembodied encounter[s] with the world' (Bærenholdt *et al.* 2004: 101). Rather, they are both produced by and give rise to, a sensual poesis as the visual finds presence through the materiality and corporeality of the body (see Scarles 2009). Therefore, while some attend primarily to the embodied performances of the tourist experience (e.g, Bærenholdt *et al.* 2004; Game 1991 and Obrador-Pons 2003, 2007), the visual exists as a series of embodied practices as tourists encounter the world multisensually and multidimensionally (Crouch and Lübbren 2003). In acknowledging the body as an active agent in the making of knowledge (Crang 2003), the visual becomes inherently implicated in and reliant upon the ways in which we taste, smell, touch and hear within and among our emergent surroundings. Thus, photographs and photography not only become implicit in the ways in which tourists produce and consume places, but also in the way in which they *communicate* such experiences.

As this book highlights, visuals can be introduced to the research setting through a variety of means, whether still images through photography or moving images via methods such as filming and the creation of video documentaries or diaries. Indeed, referring to Harper (1998), Warren (2005: 862) suggests: 'on one level all research practice is visual since we are in the business of describing researched worlds to our readers and students so that they can visualise our words'. In order to explore the ways in which the visual can mobilise an embodied expression of self this chapter first attends to the method of photo-elicitation. Moving beyond the notion of images as providing data (for example, via pre-existing archives of advertising literature, brochure images, or archival photographs documenting the development of a tourist destination) attention turns to the notion of images as *producing* data. Photographs become active agents within the research process as greater emphases lie on subjective meaning and the practices and processes behind the creation of the image; not what is represented, but what is done and why (Crang 2003). As Ruby (1995, cit. Bignante 2009) suggests, visuals provide the opportunity to explore respondents' social and personal meanings and values by their response to images. Interest therefore lies not only in the visual as object or artefact, but in the active, embodied practices and performances that underpin the significance of the visual and created the need for its being

(Radley and Taylor 2003; Scarles 2009). Visuals are therefore often combined with other techniques such as interviewing, focus groups, researcher or respondent diaries and so forth, as a means of furthering communication and opportunities for respondents to express and explore experiences of particular research phenomenon.

While the use of photographs as a technique of elicitation originated with the work of Collier (1957, 1967), early examples of such techniques in tourism research remained absent until work from authors such as Botterill and Crompton (1987) and Botterill (1988, 1989) where photo-elicitation was employed to understand tourist experience using tourists own photographs to aid discussions. Combining photo-elicitation with repertory grid techniques, Botterill and Crompton elicited deeper discussions of the destination images held by respondents. Since then, many researchers such as Cederholm (2004), Jenkins (1999), Loeffler (2004), MacKay and Couldwell (2004) and Zainuddin (2009) have adopted such an approach. Originally introduced as native image making by Wagner in 1979, some researchers (see for example MacKay and Couldwell (2004) and Garrod (2006, 2007) now refer to respondent-led photography as volunteer- or visitor-employed photography. Where researcher-led photo-elicitation focuses on introducing respondents to photographs that are pre-selected by researchers according to established research criteria (relating for example to number of photographs, content of photographs, size of photographs, style of presentation and display of photographs, etc.), respondent-led approaches provides opportunities for respondents to produce their own images before discussing their significance and meaning with the researcher. While this can of course be influenced by specific requests from researchers (e.g., number of photographs, photographs of specific scenes or contexts, photographing within a particular timeframe, etc.), such technique affords relative freedom of respondent selectivity of content inclusion and exclusion, composition and framing enabling them to convey their subjective interpretations and experiences of place. Indeed, the affordability (Brandin 2003; Garrod 2007), increasing user-friendliness of cameras and the ubiquitous nature of photography in tourism (Chalfern 1987; Sontag 1979; Haldrup and Larsen 2003) ensures respondent familiarity, comfort and confidence in photographing personal experiences and encounters for research purposes.

Empowering respondents and accessing the emotional self via the visual

Since the emergence of photo-elicitation, the advantages of introducing photographs into interview contexts have been well documented. As MacKay and Couldwell (2004: 391) suggest, respondent-led photography offers the 'potential for capturing and analyzing people's perceptions'. Photographs within interviews facilitate rapport (Collier 1967; Harper 2002); generating spaces of comfort and establishing trust (Bignante 2009) as respondents talk around photographs showing content they themselves have selected (Radley and Taylor 2003). It mobilises opportunities for increasing the clarity of cultural meaningfulness and significance of research (Harper 1984) and sharpens respondents' observation skills (Garrod 2008). Visuals can be used to prompt respondents for deeper and richer responses (Garrod 2007, 2008), thus eliciting additional information and stimulating discussion as both respondent and researcher may bring different interpretations of image-content to the conversation (MacKay and Couldwell 2004). Indeed, it is the opportunities for respondent reflexivity that photo-elicitation affords that not only facilitates rapport between respondent and researcher, but provides security and comfort as respondents reach out to touch or hold onto the images that are present within the conversation (Oliffe and Bottorff 2007).

Such methods invariably demand collaboration and cooperation between respondent and researcher. The performative nature of photo-elicitation embraces the ability of photographs to facilitate the enlivening of respondent/researcher encounters via dynamic performances that mobilise the co-construction of knowledge imparting and exchange as respondents, researchers and visuals come together to ignite deeper, more meaningful conversational exchange. However, it is the transference of 'control' to respondents that mobilises increased significance and commitment of self and subjective experience. As Stedman *et al.* (2004) suggest, as respondents construct their own photographs they can reflect upon their experience, thus opening spaces of reflexivity within which the potential for accessing the embodied knowledges of self and other emerges. In renegotiating control away from researchers, respondents are repositioned as producers and directors as photographs are taken without the presence of the researchers and in spaces that the respondents choose and hence convey as important (Bijoux and Myers 2006; Radley and Taylor 2003). Respondent-led photography therefore introduces a multiplicity of subjective interpretations. Indeed, it is the plurality of subjectivities within tourist encounters (MacKay and Couldwell 2004) as conveyed through a variety of respondent lenses that offers researchers an insight into the range of creative, innovative practices within the tourist experience.

Wang and Burris (1992) and Warren (2005) refer to such transference of power as giving 'photo-voice' to respondents. When accessing embodied spaces of touristic experience it is not merely the transference of power that is of significance, but also the effect this has as respondents are able to construct accounts of their experiences and lives in their own terms (Holloway and Valentine 2000), thus offering an insight into aspects of the research arena from which researchers would otherwise remain excluded (Bijoux and Myers 2006; Oliffe and Bottorff 2007). Photographing becomes a means of personalising knowledge exchange as photographs are brought into existence through respondent subjectivity and engagement with the research environment as *lived*. Indeed, just as photographs become imbued within the context of the research, respondents become *imbued* within the photographs that are

Practical Tips 5.1

INSTRUCTIONS FOR RESPONDENTS

- Any instructions to respondents have to be clear and you must clearly explain the research aims and objectives and which photographs respondents are required to take/share. This can be highly specific and relate to the number of photographs or to content (e.g., particular landscapes, objects or places), or alternatively, you may only wish to explain the context of the research and leave the rest up to the respondent. Whatever you decide, it must be fully understood!
- Be clear on the timeframe from introducing the research to respondents, to them taking the photographs before conducting the interviews. Will this happen all in one day or over a longer period of time? It can be up to a few weeks later in some cases. It is important that respondents know what are the commitments that they are making.
- Be prepared to supply the cameras for respondents. You may also wish to provide a copy of the photographs for respondents to retain.

taken as they commit themselves as an entirely embodied, emotional and sensual agent within the research arena. As Garrod (2007: 17) suggests, 'such approaches are concerned with a more holistic account of the human-environment relationships, including both visual and cognitive elements . . . people are not merely viewers of landscape but are situated experientially in it'. Such embodied connections to, and performances of, their surroundings directly unite respondents to the research arena as a space of lived encounters, establishing confidence in communicating experiences at a range of different levels as photographs become embodied extensions of self in the research arena.

Mobilising spaces of embodied reflexivity and reflexive performance

In order to realise the potential of photographs as providing access to embodied knowledges and experience, it is vital to consider the role of respondent reflexivity within the space of the interview (see MacKay and Couldwell 2004). The importance of the respondent does not diminish once the photographs have been taken. Rather, their presence and voice continues to be empowered in the interview. As Lury (1998) suggests, photographs become culturally fashioned extensions of the senses. Thus, photography becomes a technique through which respondents are encouraged to 'fashion their feelings and thoughts . . . and make them visible' (Radley and Taylor 2003: 80). While many authors acknowledge the role of photographs as prompts or triggers for memory (Cronin and Gale 1996; Harper 2002), it is the role of photographs as 'beacons of personal memory' (Cloke and Pawson 2008: 16) that mobilises embodied expression according to contextualised encounters as photographed. Within elicitation, photographs become co-performers as knowledges are reproduced, shared and reflected upon through active, embodied reflexive performances (Scarles 2009). Photographs reignite the immediacy of experiences and become an arena for negotiation and play as they offer respondents the opportunity to reflect and 'access previously hidden behaviours, senses, and engagements' (ibid.: 466).

The comfort photographs offer to respondents extends beyond diverting attention away from the researcher and their questions to the familiar space of, and feelings evoked by, the image presented. Rather, comfort extends to facilitating increased self-disclosure and expression of potentially more sensitive issues around sentiments, senses, emotions, feelings, values and beliefs. Indeed, photographs may also enable respondents to share lay, unwritten and unspoken knowledges that at times evade consciousness (Meyer 1991). As Langer (1957 in Warren 2002: 229) suggests:

> everybody knows that language is a very poor medium for expressing our emotional nature. It merely names certain vaguely and crudely conceived states, but fails miserably in any attempt to convey the ever-moving patterns, the ambivalences and intricacies of inner experience, the interplay of feelings with thoughts and impressions, memories and echoes of memories.

In dwelling in a world of words (Prosser 1998), our bodies and emotions are inherently framed within language, signifiers and discourses. Yet, as Bennett (2004) realises, bodies simultaneously mediate emotions and connections between subjectivities and social worlds.

Photographs therefore provide opportunity to ignite embodied reflections that extend beyond the materiality or description of the photograph. As Sontag (1979: 23) suggests, 'the ultimate wisdom of the [photograph] . . . is to say: there is the surface. Now . . . feel, intuit – what is beyond it'. It is the personal connection to the photograph that draws respondents into the body of the image and facilitates communication of underlying narratives and embodied performances. As they reconnect with that which is photographed, embodied reflexive performances reignite that which photographs cannot show and corporeal vision cannot see (Scarles 2009). Through the 'vanishing point' (Phelan 1997), respondents reignite the interior of the image as they penetrate its interiority and once again sense what the subject feels like. Therefore, as illustrated in Scarles' (2009) study on tourists' use of photography on holiday, Olivia referred to one of her photographs as conveying the 'utter silence' and 'total isolation' of the Peruvian altiplano, while Sarah reflected upon the 'sound of them (porters) in the camp, the laughing and the joking' from her trek to Ausengate (see Figures 5.1 and 5.2).

Figure 5.1 Total isolation
Source: Respondent's photograph, used with permission

Figure 5.2 Laughter at camp
Source: Respondent's photograph, used with permission

Practical Tips 5.2

INTERVIEWING WITH PHOTOGRAPHS

- Always talk using the photographs. Refer to them and point to them or pick them up as this will also reassure respondents that they can do the same.
- Remember that conversation will move beyond that which is depicted in the photograph. While it is useful to engage respondents in a discussion of what is pictured, your conversation should not stop there and you should use the photographs as a starting point. Remember, as with interviewing, new and interesting avenues of conversation will emerge that are of interest to your research. Often what is of interest is not what is directly shown in the photograph, but the supporting narratives as to *why* particular views or objects or people, etc. were chosen to be photographed, is critical. It is very important to encourage respondents to not only describe their pictures, but to engage in conversation in the issues imbued within them.
- Allow respondents time to think and remember how they were feeling during a particular experience. Remember, everyone thinks and expresses themselves differently so it is important not only to listen, but also to observe how respondents react to particular photographs and the supporting reflexive narratives.

Importance therefore lies with the supporting narratives, gestural clues, tendencies and orientations that are subsequently brought into being and expressed. Photographs become more than mere aides to conversation; they exemplify, revivify and allow expression of that which respondents *feel* is important to the research as reflective of their subjective experiences.

Including the researcher as self: mobilising intersubjective exchange via the visual

Finally, attention turns to the opportunities visuals afford for mobilising an intersubjective togetherness between respondent and researcher via visual autoethnography (Scarles 2010). Visual autoethnography

> exists as a fusion of visual elicitation and autoethnographic encounter; an opportunity for accessing and mobilising deeper, nuanced insights into embodied performances, practices and processes of the tourist experience . . . It is no longer enough to listen and respond to respondents' narratives as they emerge via elicited visuals.
>
> (ibid.: 5)

Where photo-elicitation solicits a dynamic, co-constructive collaboration between respondent and researcher, within visual autoethnography the researcher becomes more deeply situated within the research as they themselves also become researched. For example, during my own research I spent two weeks following the 'tourist trail' around Peru: visiting the key tourist sites, eating the cuisine that tourists would be eating alongside other tourists in restaurants, talking with the local people from tour guides to villagers we met along the way. Likewise, I also took photographs (like other tourists) in order to document my travels. In addition to photo-documentation, research methods complementing visual autoethnography include: reflexive field-diaries, constructing video extracts, drawing, painting, collecting of souvenirs (including postcards), etc., in order to further engage with and understand the research environment and how, in my case, it feels to be a tourist. While such data may not be directly introduced to the later interview setting while talking to respondents, it is the emergent autoethnographic knowledges created throughout such experiences that mobilises a connectedness to both the research environment but importantly, also to the respondent. Researcher subjectivity is therefore embraced as a co-constructive force of agency as the situated knowledges of both researcher and respondent as 'active agents' (Spry 2001) are enlivened as *both* engage in a series of active doings as each experience the research environment first-hand.

In embracing the multiplicities of self and other, visual autoethnography mobilises interviews as co-constructions that 'move beyond discursive productions, productions of power and the propagation of knowledge that potentially limit expressions of self and other' (Scarles 2010: 6). The presence of visuals and the subsequent discussions emanating from that-which-is-seen, mobilises a togetherness as both researcher and respondent share experiences and establish common ground upon which conversations emerge through mutually intelligible meanings of subjective encounters (Reed-Danahay 1997). Thus, visuals 'offer gateways for merging reflexive subjectivities; the bridge that connects researcher's and respondent's experiences as they emerge within the space of the interview' (Scarles 2010: 8). As Scarles and Sanderson (2007) realise, intersubjectivity mobilises a 'sharing of speech' as respondents are able to articulate the intensities of embodied performances through the visuals presented thus, 'expressing a deeper appreciation of the multiplicity of attitudes, habits, sentiments, emotions, sensibilities and preferences of tourists' experiences' (Scarles 2010: 10). Thus, as research seeks to understand the tacit, tactile and embodied remembrances of experiences as lived (Crang 2003), researchers are able to respond to and subsequently support and *understand* respondents' encounters via mutual appreciation.

As the visual grounds conversation, opportunities arise to explore not only respondents' positive or desirable experiences, but also that which causes, among other responses, sadness, pain, banality, regret or discomfort. However, mutuality through visual autoethnography should 'not assume agreement between subjectivities as disjuncture can also arise as moments of researcher

Figure 5.3 Elation at reaching the summit

Source: Research participant's photograph, used with permission

Figure 5.4 Sensations of being on the boat

Source: Research participant's photograph, used with permission

and respondent commonality are pervaded by difference as both come to the interview space with potential disparities in worldviews and belief systems' (Scarles 2010: 7). Yet, such clashes should not be feared or actively avoided. Rather, in harnessing the visual as a point of mutuality, nuances and subjective differences become open to discussion and can enrich the research exchange as respondent and researcher realise a shared commonality (e.g., the desire to travel, a love of nature, fascination with cultural differences across communities, etc.). Consequently, conversations emerge as 'a rich negotiation, sharing and mutual understanding of experience' (ibid.: 13). The visual becomes the point of shared experience, facilitating commonality while simultaneously providing individual moments of subjective reflection as both respondent and researcher reflect upon their own personal experiences that stretch beyond that which is pictured. Therefore, during a recent project exploring the ways in which tourists utilise the visual during their tourist experience, as Maggie shared her elation at reaching the mountain summit or Angela commented that 'it's not just the visual side of it, it's the smells, it's the sounds, it's the sensations of sitting on that bloody boat going up and down', I too was able to recall my own similar experiences of achievement and likewise, seasickness having experienced similar encounters too (see Figures 5.3 and 5.4).

Yet, the very nature of embodied and affectual connection to, and performances of, place demands spaces are created within the research environment where words become redundant and 'sounds of silence' emerge (Scarles 2010). Within visual autoethnography the visual can move to occupy a space that transcends the realm of representation and narrative as respondent and researcher reflexivities extend to reveal emotions and open intimacies of self as exchange moves to embrace the realms of sensate life (Smith 2001; Thrift 1999). However, during such moments, discursive discrepancies can arise as respondents become unable to express themselves using verbal or textual dialogue. Thus, 'as words fail, visual autoethnography opens the possibility of sounds of silence as visuals allow respondents to reflect upon and imaginatively reignite their embodied practices and performances of place' (Scarles 2010: 14). Orobitg-Canal (2004) too attends to the ultimate failure of words as respondents become frustrated and imprisoned by vocabulary. Indeed, on reflecting her experiences of seeing and photographing Machu Picchu for the first time, Paula commented:

> I just thought wow I am up here . . . this is mine because I am here and I can see it, wow [laughs] . . . unless you are there you just can't believe it . . . they can't capture it because its just so big and it's so vast and it's just amazing, yeah . . . it's just like the awe I guess, you just think wow, you know and you think I know I can't really capture this but I feel I have got to take it.

While some continue in their attempts to verbally convey the intensities of experience, it is the inevitable limitations of articulation that confines

Figure 5.5 Local children

Source: Research participant's photograph, used with permission

expression. As Harrison (2008: 19) suggests: 'we come to ourselves already entwined in the unfolding historicity of many such regimes that our intentions . . . our desires, action and words will never have been quite our own'. Sounds therefore arise in what is *not* said as 'silences should not be assumed as absolute quietness as respondents sit devoid of expression or communication. Rather . . . non-verbal communication generates sounds of silence as expression resonates through the visual' (Scarles 2010: 14).

Yet, moments also arise where ramblings stop and respondents become withdrawn; the reflexive remembrances and subsequent re-enlivening of haptic, affectual spaces of experience take over. Indeed, while such silences may create disjuncture and fractures in conversation, the mutuality of visual autoethnography mobilises spaces of comfort and understanding as an unspoken 'knowing' emerges between respondent and researcher as those-who-have-experienced. Body language and gestural clues come to lend meaning and significance (Angrosino and Mays Perez 2000). Thus, as aforementioned, visuals become co-performers in the space of the interview (Holm 2008; Scarles 2009); a pathway to understanding experiences 'not just as a physical setting, but an orientation, a feeling, a tendency' (Radley and Taylor 2003: 24). Indeed, the visual can mobilise reflexive performances that can launch expressions of corporeal uniqueness as emotions exceed expression in language and erupt into gesture (Elkins 1998; Mulvey 1986). Indeed, as Sarah reflected on her experiences of meeting local children in rural Peru, her emotions took over, as referring to Figure 5.5, she explained:

> these guys are laughing because they are getting balloons, fruit, pencils. He is singing me a song, they stood there and . . . they got things, they did another one, they got things, there's another picture I have with kids running down the street and I am thinking 'oh, Jesus do I have enough?' . . . but so many kids that we gave things to, I mean when we got right out into the country and we were giving them sweeties we had to show them how to unwrap that, anyway . . . [gets very upset and stops talking].

Thus, both researcher and respondent share a vulnerability of self (Scarles 2010), manifest as a corporeality of vulnerability that 'describes the inherent and continuous susceptibility of corporeal life to the unchosen and the unforeseen' (Harrison 2008: 5). Thus, as Scarles (2010: 17) suggests,

> the ultimate failure of verbal expression should not be misinterpreted as the end of communication: a hopeless dead-end from which researchers and respondents must retreat. Rather, by combining visuals with auto-ethnography, where words fail, visuals ignite and communication continues . . . [as] visual autoethnography facilitates poetic continuations that bridge the gap between the represented and the non-representable.

Practical Tips 5.3

VISUAL AUTOETHNOGRAPHY

- If using visual autoethnography it is important that you share your own experiences. However, remember, this does not mean that your experiences must match with those of the respondent. Likewise, where differences in experiences arise this should not be interpreted as one being right and the other wrong. You must realise the importance of difference and use this to prompt further discussion and insight.
- Do not always try to fill silences with words. While they can sometimes feel uncomfortable, remember that silences tell us as much and at times more about how respondents are feeling. It is important then to provide space for gesture and expression through emotion or body language. You can later reflect upon these moments and write detailed notes in your field diary once the interview has finished.
- Don't limit discussion to that which can be seen. Remember, the content of the photograph serves to prompt and trigger different directions in conversation and it is often what is not pictured that can be of particular interest!

Potential limitations and ethical considerations

As with all methods, there are not only opportunities afforded by the use of visuals to access embodied spaces of the tourist experience and it is inevitable that limitations exist in adopting such techniques. First, respondent-led photographs introduced to interviews are inevitably context-specific; illustrating particular practices and experiences in moments abstracted both spatially and temporally. Second, in raising respondent consciousness about their environment, the potential exists for artificially raising not only the voice of the respondent but also generating false memories. Respondents can therefore perform and share experiences according to selective memories as they revisit memories and experiences to suit their current identities (Gillis 1994). Third, responses can be influenced by what respondents believe researchers want to hear, or respondents can alter responses in order to present themselves in their best light. However, relating specifically to the presence of the visual within interviews, photographs can mobilise a popularisation of memory (Edensor 1998) as reflexive performances can call forth idealised imaginings. Thus, despite being constructed by respondents, supporting narratives triggered by photographs can become caricatured as 'true' memories are replaced, or remain hidden, as realities are replaced and respondents convey affinities with that pictured according to preferred imaginings and remembrances. Fourth, upon being asked to express feelings, emotions or sentiments, some respondents may also feel unsure in their ability to fully convey that which has been experienced. As aforementioned, responses can become unfocused and rambling as visuals become implicated by language and inherently bound by text. Therefore, as Bijoux and Myers (2006: 51) suggest, such practice can create content that is 'highly variable and individualistic as well as being less detailed, or different from, what the researcher might have been most interested in'. This in turn therefore raises questions about 'the status of the image and about the reasons for the selection of the subjects pictured' (Radley and Taylor 2003: 79).

Several issues may also arise with regard to respondents' willingness to photograph. First, while this chapter emphasises the importance of the practices, experiences and performances behind the photograph, as Bijoux and Myers (2006) suggest, some respondents may not have the skills required to participate or alternatively may not be willing to commit the time and effort required to produce the photographs as well as take part in an interview. Second, some respondents can feel uncomfortable sharing their photographs and at times can become self-conscious; making excuses and apologising for the quality of the images. Indeed, as one of the author's past respondents commented: 'Are you sure you want to see my photographs? I am an expert at taking pictures of my own thumb and have been causing widespread mirth and derision at my photos showing the waters of Lake Titicaca have quite a severe slope' (Peter).

Feelings of inadequacy and failure to 'live up to expectation' are not uncommon and respondents often seek researcher's approval in the misperception that their photographic skills may be judged. It is therefore important that the researcher is entirely clear about what is expected of respondents not only in terms of research-specific context, content, and the number of photographs to be taken (or brought to the interview where respondents are perhaps selecting from personal pre-existing photographs), but also in terms of the level of skill required and the emphasis on capturing personal experience rather than producing professional, aesthetically pleasing images. Indeed, in many cases it is not the stereotypical, aesthetically pleasing classic views of places that are of interest, but rather the nuanced quirks of encounter that are often omitted from popular discourse and collective interpretations of place.

However, where respondent-led photographs are used, it is very important to incorporate several key ethical considerations. The issue of image ethics has received limited attention (see Prosser 2000; Prosser *et al.* 2008) and while there are a series of ethical guidelines that can be followed, the lack of a specific universal ethical measure for the use of images in research generates a range of interpretations and opinions of the nature and effectiveness of such measures. However, as Prosser *et al.* (2008: 18) suggest, 'where visual data is being used purely for elicitation purposes then issues of content are relatively unproblematic. However, if researchers wish to include these photos in dissemination of the research then some particular issues of consent emerge'. Generally, the issue of obtaining copyright for the use of images is easily overcome by either asking respondents to sign an image consent form where they are identifiable in images taken for the project, or alternatively asking them to provide written permission in the form of a letter where they hold copyright. However, where respondent photographs identify individuals not known to the respondent or researcher, a universally accepted ethical standpoint becomes less clear. Indeed, while privacy laws exist to protect against intrusion into an individuals personal space, it remains legal to photograph someone in a public space (see Gross *et al.* 1988; Lester 1996). Nevertheless, the ambiguity of such legality compounds confusion over rights of privacy; where and whom it is appropriate to photograph. Therefore, where respondents photograph general public scenes that do not explicitly identify individuals, researchers may choose to simply present the photograph in its original format. However, where image content is potentially sensitive researchers may wish to conceal identities (where informed consent has not been obtained) by blurring peoples faces using pixilation techniques or blackening identifiable features. It is therefore vital that researchers consider:

> the implications of what images they might be presented with by study participants and brief them about seeking permission and explaining the purpose prior to taking images of others. In some cases this may be all that is required but researchers are advised to be circumspect in the use of images of identifiable others and to consider whether or not someone might be at risk of harm or moral criticism as a result of the use of the image.
>
> (Prosser *et al.* 2008: 19)

Practical Tips 5.4

REASSURING RESPONDENTS

- Be sure respondents know exactly what is being asked of them. What photographs are they taking and why? How many should they be taking?, etc. Make sure your instructions are clear and that respondents have the opportunity to ask questions to clarify any misunderstandings they may have.
- To alleviate possible doubts, reassure respondents that the aesthetic and compositional quality of the photographs is not what is important but rather the reasons for taking the photographs.
- When talking to respondents about their photographs and experiences, remember to talk about that which is not pictured. Reassure respondents that there are no right or wrong answers, but that you are interested in hearing their own experiences.
- You should always obtain copyright permission from the respondents before using the images they generate in any published media.

CHAPTER SUMMARY

This chapter has addressed the role of the visual, in particular the introduction of respondent-led photographs to the interview setting, as a means of accessing embodied, performative spaces of the tourist experience. The following key conclusions can be drawn from this chapter:

- The visual can become more than a mere aide-memoire that elicits responses. By understanding respondent-led photography as directly implicated in, and constructed through the ways in which we taste, smell, touch, hear as well as see the world, photographs become implicit in the ways in which respondents (as tourists) both produce and consume place. The visual can therefore access the nuanced moments of the tourist experience that come to exist within the embodied, haptic and affective spaces of encounter between self and other.
- Visuals become active agents and co-performers in the research process. As Crang (2003) and Radley and Taylor (2003) realise, unlike content or semiotic analysis, importance lies not with the content of the photograph per se, but rather the circumstances that have created the need for its being.
- Empowering respondents in the research process is vital to the successful application of the visual in research methods. Unlike researcher-led photography, respondent-led photography transfers control to respondents; spaces are opened for respondent reflexivity as they are repositioned as producers, creators and directors of the experiences to be communicated.

- Through visual autoethnography deeper intersubjectivity between respondent and researcher emerges. Visuals act as bridges between respondent and researcher experiences (Scarles 2010); mobilising togetherness as researcher and respondent establish common ground and conversations emerge through mutually intelligible meanings of subjective experiences (Reed-Danahay, 1997).
- Through empathy and *understanding*, spaces of mutuality emerge. Where words fail, visuals ignite conversations as a sharing of speech and sounds of silence emerge (Scarles 2010). The visual should therefore not only be understood as that which can be seen, but rather as that which is *lived*; expressed via a fusion of *all* our senses as researcher and respondent come together through spaces of understanding and a desire to know.

Annotated further reading

Pink, S. (2007) *The Future of Visual Anthropology: Engaging the Senses*. London: Routledge.

Sarah Pink reconceptualises our understanding of the visual in this text as she readdresses the visual as a sensual medium through which researchers can engage with the world. In doing so, she presents a range of conceptual understandings of the visual from sensual engagement to social intervention and accessing research spaces through hypermedia.

Radley, A., and Taylor, D. (2003) 'Images of Recovery: Photo-elicitation Study on the Hospital Ward', *Qualitative Health Research*, 13: 77–99.

This article offers an informative study that not only address the theory of visuals in interviewing, but offers an applied insight into the ways in which photographs can be used to access the sensual spaces of organisational life.

Rose, G. (2001) *Visual Methodologies: An Introduction to the Interpretation of Visual Materials*. London: Sage.

This text offers a comprehensive insight into the range of visual methods available to researchers. Although not specifically aimed at tourism, this text outlines a series of visual methods as forms of interpretation and analysis. Additionally, it provides a critical contextualisation of the visual as a research tool and method.

Scarles, C. (2009) 'Becoming Tourist: Renegotiating the Visual in the Tourist Experience', *Environment and Planning D: Society and Space*, 27: 65–488.

This publication discusses the importance of understanding the visual as a series of practices and performances that extend beyond ocularcentrism and embrace the visual as entirely embodied and affectual in nature. As a multisensual encounter, the visual (in particular photography) is presented as a series of complex performative spaces that permeate the entire tourist experience.

Scarles, C. (2010) 'Where Words Fail, Visuals Ignite', *Annals of Tourism Research*, 37: 905–26.

This newly published article by Caroline Scarles offers an exploration of the opportunities of visual autoethnography in tourism research. The article extends thinking to embrace the opportunities of mobilising intersubjectivity between both researcher and respondent as a means of accessing the embodied spaces of tourism research.

Warren, S. (2002) 'Show Me How It Feels To Work Here: Using Photography to Research Organisational Aesthetics', *Ephemera: Critical Dialogues on Organisation*, 2: 224–45.

Similar to the publication above, this article also contains a study that addresses both the theory of visuals in interviewing and offers an applied insight into the ways in which photographs can be used to access the sensual spaces of organisational life.

References

Adler, J. (1989) 'Origins of Sightseeing', *Annals of Tourism Research*, 16: 7–29.

Angrosino, M., and Mays Perez. K. A. (2000) 'Rethinking Observation: From Method to Context', in N. K. Denzin and Y. S. Lincoln (eds) *Handbook of Qualitative Research*, London: Sage.

Bærenholdt, J. O., Haldrup, M., Larsen, J., and Urry, J. (2004) *Performing Tourist Places: New Directions in Tourism Analysis*. Aldershot: Ashgate.

Bennett. K. (2004) 'Emotionally Intelligent Research', *Area*, 36: 414–22.

Bignante, E. (2009) 'The Use of Photo-Elicitation in Field Research: Exploring Maasai Representations and Use of Natural Resources', *EchoGéo*, 1: 11–16.

Bijoux, D., and Myers, J. (2006) 'Interviews, Solicited Diaries and Photography: "New" Ways of Accessing Everyday Experiences of Place', *Graduate Journal of Asia-Pacific Studies*, 4: 44–64.

Botterill, T. D. (1988) *Experiencing Vacations: Personal Construct Psychology, The Contemporary Tourist and The Photographic Image*. Unpublished Ph.D. Thesis. Texas A and M University: Texas.

Botterill, T. D. (1989). 'Humanistic Tourism? Personal Constructions of a Tourist: Sam Revisits Japan', *Leisure Studies*, 8: 281–93.

Botterill, T. D., and Crompton, J. L. (1987) 'Personal Constructions of Holiday Snapshots', *Annals of Tourism Research*, 14: 152–6.

Brandin, E. (2003) 'Disposable Camera Snapshots: A Method of Interviewing Tourists', Presented at Tourism and Photography: Still Visions – Changing Lives Conference, Sheffield Hallam University, United Kingdom, 20–23 July.

Cederholm, E. A. (2004) 'The Use of Photo-Elicitation in Tourism Research: Framing The Backpacker Experience', *Scandinavian Journal of Hospitality and Tourism*, 4: 225–41.

Chalfern, R. (1987) *Snapshot Versions of Life*. Bowling Green, OH: Popular Press.

Cloke, P., and Pawson, E. (2008) 'Memorial Trees and Treescape Memories', *Environment and Planning D: Society and Space*, 26: 107–22.

Cohen, E. (1988) 'Authenticity and Commodification in Tourism', *Annals of Tourism Research*, 15: 371–86.

Coleman, S., and Crang, M. (2002) *Tourism: Between Place and Performance*, Oxford: Blackwells Publishing.

Collier. J. (1957) 'Photography in Anthropology: A Report on Two Experiments', *American Anthropologist*, 59: 843–59.

Collier, J. (1967) *Visual Anthropology: Photography as Research Method.* New York: Holt, Rinehart and Winston.

Craik, J. (1997) 'The Culture of Tourism', in C. Rojek and J. Urry (eds) *Touring Cultures: Transformations of Travel and Theory*. London: Routledge.

Crang, M. (1997) 'Picturing Practices: Research Through the Tourist Gaze', *Progress in Human Geography*, 21: 359–73.
Crang, M. (1999) 'Knowing, Tourism and Practices of Vision', in D. Crouch (ed.) *Leisure/Tourism Geographies: Practices and Geographical Knowledge*. London: Routledge.
Crang, M. (2002) 'Qualitative Methods: The New Orthodoxy?', *Progress in Human Geography*, 26: 647–55.
Crang, M. (2003) 'Qualitative Methods: Touchy, Feeling, Look-see?', *Progress in Human Geography*, 27: 494–504.
Cronin, O., and Gale, T. (1996) 'Photographs and Therapeutic Process', *Clinical Psychology Forum*, 89: 24–8.
Crouch, D. (2000a) 'Introduction: Encounters in Leisure/Tourism', in D. Crouch (ed.) *Leisure/Tourism Geographies: Practices and Geographical Knowledge*. London: Routledge.
Crouch, D. (2000b) 'Places Around Us: Embodied Lay Geographies in Leisure and Tourism', *Leisure Studies*, 19: 63–76.
Crouch, D., and Lübbren, N. (2003) *Visual Culture and Tourism*. Oxford: Berg.
Dann, G. M. (1988) 'Images of Cyprus Projected by Tour Operators', *Problems of Tourism*, 41: 43–70.
Dilley, R. (1986) 'Tourist Brochures and Tourist Images', *Canadian Geographer*, 31: 59–65.
Edensor, T. (1998). *Tourists at the Taj: Performance and Meaning at a Symbolic Site*. London: Routledge.
Edensor, T. (2000) 'Staging Tourism: Tourists as Performers', *Annals of Tourism Research*, 7: 322–44.
Edensor, T. (2001) 'Performing Tourism: Staging Tourism: (Re)Producing Tourist Space and Practice', *Tourist Studies*, 1: 59–81.
Edwards, E. (1996) 'Postcards: Greeting from Another World', in Selwyn T. (ed.) *The Tourist Image: Myth and Myth-making in Tourism*. Chichester: Wiley.
Elkins, J. (1998) *On Pictures and Words That Fail Them*. Cambridge: Cambridge University Press.
Franklin, A. (2003) *Tourism: An Introduction*. London: Sage.
Franklin, A., and Crang, M. (2001) 'The Trouble With Travel and Tourism Theory', *Tourist Studies*, 1: 5–22.
Game, E. (1991) *Undoing the Social: Towards a Deconstructive Sociology*. Toronto: University of Toronto Press.
Garrod, B. (2006) '"It was quite nice here before the tourists came": Using Volunteer-employed Photography (VEP) to Explore Tensions between Residents and Tourists in a Welsh Seaside Town', *The Rural Citizen: Governance, Culture and Wellbeing in the 21st Century*, UK: University of Plymouth Compilation.
Garrod, B. (2007) 'A Snapshot into the Past: The Utility of Volunteer-employed Photography in Planning and Managing Heritage Tourism', *Journal of Heritage Tourism*, 2(1): 14–35.
Garrod, B. (2008) 'Exploring Place Perception: A Photo-based Analysis', *Annals of Tourism Research*, 35: 381–401.
Gillis, J. R. (1994) 'Memory and Identity: The History of A Relationship', in J. R. Gillis (ed.) *Commemorations: The Politics of National Identity*. Chichester: Princeton University Press.

Graburn, N. H. H. (1977) 'Tourism: The Sacred Journey', in V. L. Smith (1978) (ed.) *Hosts and Guests: The Anthropology of Tourism*. Oxford: Blackwell Publishing.
Gross, L. P., Katz, J. S., and Ruby, J. (1988) 'Introduction: A Moral Pause', in L. P. Gross, J. S. Katz and J. Ruby (eds) *Image Ethics: The Moral Rights of Subjects in Photographs, Film and Television*. Oxford: Oxford University Press.
Haldrup, M., and Larsen, J. (2003) 'The Family Gaze', *Tourist Studies*, 3: 23–46.
Harper, D. (1984) 'Meaning and Work: A Study in Photo Elicitation', *International Journal of Visual Sociology*, 2: 20–43.
Harper, D. (2002) 'Talking About Pictures: A Case for Photo-elicitation', *Visual Studies*, 17: 13–26.
Harrison, P. (2008) 'Corporeal Remains: Vulnerability, Proximity, and Living on After the End of the World', *Environment and Planning A*, 40: 423–45.
Holm, G. (2008) 'Photography as a Performance', *Forum: Qualitative Social Research*, 9(2), retrieved on 12 June 2009 from www.qualitative-research.net/index.php/fqs/article/viewArticle/394/856.
Holloway, S. L., and Valentine, G. (2000) 'Spatiality and the New Social Studies of Childhood', *Sociology*, 34: 763–83.
Jenkins, O. H. (1999) 'Understanding and Measuring Tourist Destination Images', *International Journal of Tourism Research*, 1: 1–15.
Jay, M. (1997) *Downcast Eyes: The Denigration of Vision in Twentieth-Century French Thought*. Berkeley: University of California Press.
Lester, P. M. (1996) 'Photojournalism Ethics: Timeless Issues', in M. Emery and T. Curtis (eds) *Readings in Mass Communications*. Dubuque: Brown & Benchmark.
Loeffler, T. A. (2004) 'A Photo Elicitation Study of the Meanings of Outdoor Adventure Experiences', *Journal of Leisure Research*, 36: 536–56.
Löfgren, O. (1999) *On Holiday: A History of Vacationing*. London: University of California Press.
Lury, C. (1998) *Prosthetic Culture: Photography, Memory and Identity*. London: Routledge.
MacCannell, D. (1973) 'Staged Authenticity: Arrangements of Social Space in Tourist Settings', *American Journal of Sociology*, 79: 589–603.
MacKay, K. J., and Couldwell, C. M. (2004) 'Using Visitor-employed Photography to Investigate Destination Image', *Journal of Travel Research*, 42: 390–96.
Markwick, M. (2001). 'Postcards from Malta: Image, Consumption, Context', *Annals of Tourism Research*, 28: 417–38.
Meyer, A. D. (1991) 'Visual Data in Organisational Research', *Organisation Science*, 2: 218–36.
Mulvey, L. (1986) *On Visual and Other Pleasures*. London: Palgrave.
Obrador-Pons, P. (2003) 'Being-on-holiday: Tourist Dwelling, Bodies and Place', *Tourist Studies*, 3: 47–66.
Obrador-Pons, P. (2007) 'A Haptic Geography of the Beach: Naked Bodies, Vision and Touch', *Social and Cultural Geography*, 8: 123–41.
Oliffe, J. L., and Bottorff, J. L. (2007) 'Further Than The Eye Can See? Photo Elicitation and Research With Men', *Qualitative Health Research*, 17: 850–58.
Orobitg-Canal, G. (2004) 'Photography in the Field: Work and Image in Ethnographic Research', in S. Pink, L. Kürti and A. I. Afonso (eds) *Working Images: Visual Research and Representation in Ethnography*. London: Routledge.
Phelan, P. (1997) *Mourning Sex: Performing Public Memories*. London: Routledge.
Pink, S. (2007) *The Future of Visual Anthropology: Engaging the Senses*. London: Routledge.

Prosser, J. (ed.) (1998) *Image-Based Research: A Sourcebook for Qualitative Researchers*. London: Falmer.

Prosser, J. (2000) 'The Moral Maze of Image Ethics', in H. Simons and R. Usher (eds) *Situated Ethics*. London: Routledge.

Radley, A., and Taylor, D. (2003) 'Images of Recovery: Photo-elicitation Study on the Hospital Ward', *Qualitative Health Research*, 13: 77–99.

Reed-Danahay, D. E. (1997) *Auto/Ethnography*. Oxford: Berg.

Rojek, C., and Urry, J. (1997) 'Transformations in Travel and Theory', in C. Rojek and J. Urry (eds) *Touring Cultures: Transformations in Travel and Theory*. London: Routledge.

Rose, G., and Gregson, N. (2000) 'Taking Butler Elsewhere: Performativities, Spatialities and Subjectivities', *Environment and Planning D: Society and Space*, 18: 433–52.

Scarles, C. (2009) 'Becoming Tourist: Renegotiating the Visual in the Tourist Experience', *Environment and Planning D: Society and Space*, 27: 465–88.

Scarles, C. (2010) 'Where Words Fail, Visuals Ignite', *Annals of Tourism Research*, 37: 905–26.

Scarles, C., and Sanderson, E. (2007) 'Becoming Researched: Opportunities for Autoethnography in the Field', TTRA Annual Conference, Las Vegas, 17–20 June.

Smith, S. J. (2001) 'Doing Qualitative Research: From Interpretation to Action' in M. Limb and C. Dwyer (eds) *Qualitative Methodologies for Geographers*, London: Arnold.

Spinney, J. (2006) 'A Place of Sense: A Kinaesthetic Ethnography of Cyclists on Mont Ventoux', *Environment and Planning D: Society and Space*, 24: 709–32.

Sontag, S. (1979) *On Photography*. London: Penguin.

Spry, T. (2001) 'Performing Autoethnography: An Embodied Methodological Praxis', *Qualitative Inquiry*, 7: 706–32.

Stedman, R., Beckley, T., Wallace, S., and Ambard, M. (2004). 'A Picture *and* 1000 Words: Using Resident-employed Photography to Understand Attachment to High Amenity Places', *Journal of Leisure Research*, 36: 580–606.

Thrift, N. (1999) 'Steps to An Ecology of Place', in Massey, D., Allen, J., and Sarre, P (eds) *Human Geography Today*. London: Polity Press.

Thurot, J., and Thurot, G. (1983) 'The Ideology of Class and Tourism: Confronting The Discourse of Advertising', *Annals of Tourism Research*, 10: 173–89.

Tribe, J. (2004) 'Knowing About Tourism: Epistemological Issues', in J. Phillimore and L. Goodson (eds) *Qualitative Research in Tourism: Ontologies, Epistemologies and Methodologies*. London: Routledge.

Veijola, S., and Jokinen, E. (1994) 'The Body in Tourism', *Theory, Culture and Society*, 6: 125–51.

Wang, C., and Burris, M. A. (1992) 'Photovoice: Concept, Methodology, and Use for Participatory Needs Assessment', *Health Education and Behaviour*, 24: 369–87.

Warren, S. (2002) 'Show Me How It Feels To Work Here': Using Photography to Research Organisational Aesthetics', *Ephemera: Critical Dialogues on Organisation*, 2: 224–45.

Warren, S. (2005) 'Photography and Voice in Critical Qualitative Management Research', *Accounting, Auditing and Accountability Journal*, 18: 861–82.

Urry, J. (1990) *The Tourist Gaze*. London: Sage Publishing.

Zainuddin, A. H. (2009) 'Using Photo Elicitation in Identifying Tourist Motivational Attributes for Visiting Taman Negara, Malaysia', *Management Science and Engineering*, 3: 9–16.

6 Photo-elicitation and the construction of tourist experiences

Photographs as mediators in interviews

Erika Andersson Cederholm

Introduction

As indicated by Scarles in Chapter 5, the technique of photo-elicitation has been presented as a method to elicit responses in interviews with photographs as a device (Prosser and Schwartz 1998). Used in disciplines such as sociology and anthropology, it has been described as an efficient method to get close to the participants, and to discover the subject's own categorisations and definitions of his/her life-world (Harper 1988). Consequently, the method of photo-elicitation has often been portrayed as a 'can-opener' technique in the social situation of the interview. However, in tourism, where taking, making, editing and displaying photographs is an integral part of the subject's life-world, visual images play a more important role than simply being a convenient device or 'can-opener'. This chapter will illuminate the multifaceted potential of photo-elicitation techniques in tourism research by providing an opportunity to see this specific method as a gateway to discussions of how to understand and analyse tourist experiences. In line with the overall aim of this book, and furthering the discussions in Chapter 5, this chapter will present, discuss and promote the use of photo-elicitation as a data-collection technique for tourism researchers, but also suggest how photo-elicitation may be used for a practice-oriented narrative analysis. Through two different Swedish studies that both focus on the construction of tourist experiences through narratives, the chapter will present a social constructionist perspective on tourism research that illuminates the analytical potential of the interview situation and the practices surrounding the display of photographs.

The chapter will start with a short introduction of a practice-oriented narrative perspective and the two Swedish studies, followed by a more general overview on the collaborative situation between researcher and participants in interviews with photographs. Here, notions of authority related to researchers and participants in different forms of photo-elicitation techniques are discussed, as well as the context of production and display, which is important to consider in the analysis. The next section discusses how photographs may have the role of mediators rather than elicitors in the interview, and examples of four analytical categories are presented – 'typifications', 'the non-typical photograph', 'ambiguity' and 'social closeness and distance' – that highlight the importance of the interactive context of an interview or observation in order to analyse tourist experiences.

A practice-oriented narrative analysis and two Swedish examples

The concept of narrative in photo-elicitation research implies that the focus of the analysis is on the conversations, narratives and accounts that emerge through the interactive situation of the interview or observation. It is thus not the content of the photographs that is in focus, but the conversation and practices that accompany and interact with the photographic images. It implies that it is the participants' situated interpretations of the photographs that are of interest in this case. My approach thus differs from the realist tradition that has dominated much visual sociology in the past, where photographs are seen as reflections of reality and consequently, the studies have been much content-oriented (Pink 2001). Rather, I will present a social constructionist approach to the method of photo-elicitation. Since the researcher might be regarded as a participant observer of the interview, s/he is also part of a co-creating situation, where a social reality is collaboratively constructed by all participants involved. The approach to the photo-elicitation method that I will illustrate is an example of what Holstein and Gubrium (1997) call 'active interviewing', where the specific social situation of the interview or observation is considered in the analysis, not merely as something that may 'influence' the data-collection in a positive or negative way. According to this line of thought, I want to emphasise that the narratives and accounts that constitute the interview are not merely words but practices as well. How people say things, that is, how you do things with words, as much as the content, may be crucial for the analysis. Furthermore, the display of photographs involves specific practices: the participants point at them, touch them and order them. Furthermore, they may also deal with technical equipment such as a computer or projector, which highlights the materiality involved in interviews with photographs.

With examples from my own research, I will show how tourists as well as people working with tourism talk about and relate to photographs in individual interviews. I will also give examples from observations of photo-shows, where tourists show pictures and convey their experiences to an audience. These are situations that although taking place afterwards rather than during the trip, they nevertheless involve the active construction of tourist experiences. My examples are chosen from two different studies in Sweden. One is a study of Swedish backpackers where photo-elicited interviews with homecoming travellers were conducted together with observations from photo-shows among members of a backpacker club. The other is a study on small-scale rural entrepreneurs in the tourism and hospitality business, whose enterprises (Bed and Breakfast, horse-riding tours, galleries and cafés etc.) have a clear lifestyle orientation. This implies that for these entrepreneurs, the businesses are not run predominantly for economic profit, but as a way of making a living while pursuing their own lifestyle interests (Andersson Cederholm and Hultman 2010). Thus, the notion of gaining interesting, exciting, pleasant and novel experiences through everyday work and interactions with the guests/visitors, are as important for them as providing experiences for the guests. This convergence between the aims of the service provider and those of the tourists was particularly highlighted in one case, where the owner of a B&B has taken photographs of her guests. Furthermore, as producers of tourism services, these entrepreneurs are often dealing with visual images such as brochures and websites, which are often pointed at and referred to in the interviews. How the tourist experience – either the participants' own and/or the experiences they attempt to provide for others – is narrated and socially constructed has thus been an important analytical theme in both of these studies and, as I will show, they provide examples of how photographs and practices surrounding photographs function as an important mediator in tourism narratives. Hence, photo-elicitation in tourism studies is not merely a technique for data-collection, but provides an important basis for analysing tourism practices and experiences.

Photo-elicited interviews and conversations: a collaborative situation

Authority

In line with a social constructionist perspective, an interview or an observation is always a collaborative situation. All participants involved, passive or active, contribute to the construction of data. However, the issue of authority, especially the relationship between the researcher and participant/s, is interesting in photo-elicitation, since the researcher's authority is often altered or called into question, particularly if the participant brings his/her own photographs or is the one who chooses the images that guide the interview. Photo-elicitation is thus a reflexive mode of researching (Harper 1988: 65), where the participant, rather than the researcher, is the 'expert' guiding the interview (Collier 1967). In that sense, the collaborative dimension between the researcher and the subject may be more highlighted in interviews with photographs than oral-only interviews (see Banks 2001:96).

However, the level of authority connected to the role of researcher and researched is related to the type of interview chosen. First, we have interviews where the photographs are taken and/or chosen by the researcher. The images may be archival photographs taken by someone else and more or less accessible to the researcher and the participant alike, or the photographs may be taken by the researcher her/himself for the purpose of the research project (see Schwartz 1989). The researcher may, for instance, take photographs or film everyday situations, and then use the still photographs or film sequences

Practical Tips 6.1

RESEARCHER-PRODUCED/INTRODUCED PHOTOGRAPHS

The technique of researcher produced/introduced photographs will:

- provide more authority to the researcher and thus more control over the interview process;
- provide a certain authority to the participant when he/she chooses photographs/topics of conversation;
- possibly produce less in-depth data of the participants' life world due to the researcher-controlled situation;
- encourage the participant to reflect upon otherwise taken for granted everyday situations and behaviour.

in the interview situation. It allows the participant to reflect upon everyday, often non-reflected behaviour, and talk about it (see Heisley and Levy 1991). This type of technique may be useful in order to make the interviewee distance him/herself from daily routines and his/her own behaviour, and may thus evoke reflection. Authority in those instances may be more related to the selection of photographs: whether the researcher has selected images before the interview or if it is up to the participant to select them during the course of the interview. In those instances when the researcher has photographed or filmed the participant and his/her everyday life, the concept 'autodriven' is sometimes used in order to emphasise the participant's authority and that the response is directly related to the participant's own life (ibid.).

Second, we have the participant-produced photographs. The term 'autodriven' may be used in those instances as well (Samuels 2004) or the term 'photo-voice' has been used to acknowledge that it is the participant who is the author of the photographs (Oliffe and Bottorff 2007). Participants may be requested by the researcher to take photographs and then bring to the interview, or the researcher may ask the participants to bring photographs produced in 'natural' situations, such as family albums (Walker and Moulton 1989) or travel photography (Andersson Cederholm 2004). On some occasions, the use of photographs in the interview is not planned by the researcher, but occurs spontaneously during the interview. In my study of Swedish backpackers (Andersson Cederholm 1999), the first interviews were conducted without photographs, but since several of my participants were eager to show the pictures they had taken while they were talking about their experiences, the use of photo-elicitation was developed during the course of the study. I realised that their stories about experiences were 'visualised' to a large extent, and talking about these memories while showing the photographs came naturally to the travellers. Comments like 'I will show you the pictures, just a minute' were common and the sit-down interview was often broken up when the

Practical Tips 6.2

PARTICIPANT-PRODUCED/INTRODUCED PHOTOGRAPHS

The technique of participant produced/introduced photograph will:

- provide more authority to the participant and control over the interview process;
- provide a certain authority to the researcher when the participant is requested to take photographs for the purpose of the research project;
- occasionally encourage spontaneous photo-elicitation, which may provide important insights in the role of photography in the participant's life;
- provide insights in how tourist experiences are 'visualised'.

participant started to walk around in the house, trying to find the photographs/ album or equipment to show them. This 'break' proved to be useful though, since it gave more time and therefore more authority to the participant, which was in line with my purpose to capture the traveller's own way of narrating and framing experiences. Several of these interviews came to resemble the homecoming ritual of photo-shows with family and friends as an audience, which leads us to the issue of where and how the photographs are produced, edited and displayed.

The context of production and display

Photographs are culturally and socially embedded and, as Marcus Banks (2001: 79) points out, many of the visual forms that researchers deal with are multiple embedded. For instance, the context in which the travel photography was produced, the context in which the pictures are edited, and the context of display may all be relevant for the research project. This is particularly pertinent in tourism, since practicing tourism, being a tourist or being involved in the service work in the tourism business, implies practices related to photographs, either by taking those photographs yourself, editing them and showing them, or producing visual images to be used in various media and brochures, either for business or educational purposes.

The practices of dealing with and relating to visual images are various. They are sometimes structured in a ritual form, such as taking photographs on specific sites or occasions, or displaying them in the homecoming ritual of photo-shows for friends and family. Although these practices may be demarcated as specific events, many of the activities that involve photography are routine-based and therefore often taken-for-granted activities in tourism. Although being a tourist may be perceived as an extraordinary, non-everyday type of activity, it is in this case still relevant to think about these practices as 'everyday' since being a tourist involves ritualistic and often routine-like practices. Observations of tourists practicing 'everyday' tourism may, for instance, include tourists taking photographs. In this chapter though, I want to pay attention to the post-trip and often taken-for-granted practices related to tourist photography. Whereas some practices are ritually structured and framed as specific events, others are merely integrated in the flow of everyday life. Observing specific events, such as my observations of photo-shows organised by a backpacker club is one method I have mentioned. When it comes to interviews, especially an un-structured in-depth interview, the situation may resemble a photo-show for families and friends, but it is still framed as an interview and thus a different type of situation than a sociable family ritual. The context of the interview or observation, which is also the context of display, is important to consider not merely because some situations make it easier or harder for the researcher to have glimpses into the social and cultural world of the participants, but the collaborative interactive situation of the interview/observation actually constructs the data that is emerging (see Rapley 2001).

One factor that influences the context of the interview is whether the photographs are taken by the researcher or the participant. Opposed to the type of studies mentioned before, where photographs taken by the researcher may evoke reflections concerning the participant's own behaviour (Heisley and Levy 1991), participant-produced photographs may not evoke the same kind of reflections. Tourist photographs, as well as family photographs in general, are often composed according to specific conventions of a 'happy occasion', 'beautiful scenery' or 'unique attraction', which makes them relatively predictable and thus not as useful as a device for reflection, or at least a certain kind of reflection. There are, however, exceptions or 'deviating photographs' that may provide the interview with an interesting turn (see the section on the

Practical Tips 6.3

CONTEXT OF PRODUCTION AND DISPLAY IN THE ANALYSIS

The following dimensions of the context of production and display are important to consider in the analysis:

- Whether these contexts are socially and culturally demarcated as specific events, or embedded in the flow of everyday life.
- To what extent does the interview situation resemble or differ from other non-research based contexts of photographic display?
- In what way are the emotional responses evoked by the interview situation related to cultural and social expectations of tourism and travelling as 'joyful' occasions?

non-typical photograph, p. 100). Likewise, some of the advantages of photo-elicited interviews compared to oral-only interviews, are sometimes pointed out as having a greater possibility for emotional response, since the photographs and the situation in which they are displayed may trigger emotional responses (Collier 1967; Samuels 2004). As with the issue of reflection mentioned earlier, these emotional responses may be related to whether the photographs are taken by the researcher and thus have a surprise dimension, or if they are taken by the participant him/herself. In the case of tourism, photographs and the context in which they are displayed are often assumed to be associated with joyful events of togetherness or exciting experiences that may structure emotional responses in a certain way.

The role of the photograph – from eliciting to mediating

The concept of 'eliciting' implies getting access to otherwise concealed and difficult-to-convey perspectives of the participant's life-world (see Pink 2001: 68). Opposed to the 'can-opening' approach mentioned in the introduction, I would like to emphasise the photograph's role as a mediator with an active role in the interactive situation. The photograph might be ascribed a certain agency implying that it is not merely a passive artefact but an object of significant meaning that acts upon the social context as well as being acted upon. This is in line with the practice-oriented approach aforementioned (see also Schatzki *et al.* 2001), since the participants of the interview or observation interact with the photographs and related artefacts (computer, projector or album) in a manner that frames and structures the situation. Banks, for instance, observed that the materiality of eliciting artefacts, in his example, a television in the room where he conducted the interview, might play a significant role and become embodied agents. The television, according to Banks, became a literal voice in the conversation, 'like an elderly relative sitting in the corner mumbling away to themselves and occasionally saying something interesting' (Banks 2001: 86). Although a television set may have more 'voice' than a still photograph, the photograph is part of an interaction and it is thus relevant for the analysis what people do with the photographs or other artefacts while they are talking. In this sense, the interview and the observation should not be considered as two disparate forms of data collection method, but as sites of interaction that may involve more or less talk and non-verbal practices. Thus, it is important to see the interview as a form of observation, implying that it is not only interesting what the respondents say, but how they say it (Gubrium and Holstein 1997) and what they actually do while saying it.

Oliffe and Bottorff (2007) for example, used photo-elicitation techniques in order to capture men's experiences of prostate cancer treatment and the life thereafter. With the aim of challenging common gender assumptions about older Anglo-Australian men's stoicism (a group from which the study participants were recruited) and reluctance to talk in detail about prostate cancer issues such as incontinence and sexual dysfunction, the use of photo-elicitation seemed to evoke openness and emotional responses from the men in the study that challenged such gender expectations. However, what they also found was that the practices of taking and showing pictures evoked informal interactions between researchers and participants, such as when one of the men gave a guided tour of his shed which had appeared on his photographs. Such interactions facilitated and legitimised 'men's talk' and 'show-and-tell' activities in and around the actual interviews, and the men could in that way retain a sense of control and autonomy even though they were talking about highly emotional and intimate issues (ibid.: 852–3). According to the authors, these men were thus actually 'doing gender' during the course of the interview. The gender aspect became visible in a new way, not merely by what the men were talking about, but by the manner in which they were doing things while they were talking.

The doing-aspect of the interview may be particularly important in narrative analysis of tourist experiences. The social construction of the tourist experience often takes place through interactions with the photographs, partly at the moment of taking the photographs in the act of photographing, partly afterwards, in the situation of remembrance and narrating. Thus, photographs do not merely elicit responses and trigger memories of past experiences, but are important objects in a specific social context – the interview/observation situation – where experiences are narrated, constructed and thus experienced (albeit differently than the past travel experiences). I will show interview/observation contexts that have resulted in four analytical categories that highlight the importance of the social, interactive context of the interview and observation in order to study tourist experiences: 'Typifications', 'the non-typical photograph', 'ambiguity' and 'social closeness and distance'.

Typifications: from individual responses to the social construction of experiences

The two studies I use as examples in this chapter – the study of backpackers and lifestyle entrepreneurs – are different in many ways, but the social construction of experiences is a common analytical denominator. Although the contexts of experiences are different, the modes in which they are constructed and the role visual images play in these contexts have similarities. Particularly the backpacker photo-shows, which are relatively ritualised and public situations, and the situation in which the lifestyle entrepreneur talks about experiences that the tourist can expect in a brochure-like manner, as when they refer to their own brochures. For example, such as 'Look here, we have a beautiful surrounding here' and 'We have different kinds of activities', highlight experiences that are to a great extent 'typified'. The manner in which they talk about the experiences in these situations are more or less according to a social norm that is reinforced in the situation, on what type of experiences are expected, and what type of experiences are socially constructed as 'experiences worth having'. In these interactive situations with visual images produced by the participants, the image and the way the participant/s interact with the photograph/s, frame the situation as a relatively predictable 'conveying the expected experience' occasion. The situation is relatively formal and the participants do not expect too much deviation. As a researcher participating in the photo-show, I would be very surprised if the traveller who is presenting will tell about his/her boring and non-educating travelling experiences to places of no interest. I would be equally surprised if the B&B owner deviates from the language of the brochure, when she explains the attractive surroundings, things to do and to see. Then, would an interview with the B&B owner without the mediating visual images convey a different picture? Or would an observation of a situation where backpackers tell stories to each other without the presence of photographs produce totally different kind of data? Not necessarily, but the presence of the photographs will probably reinforce the typification process, illustrated through the eagerness by which the participants reach for pictures when the typical will be conveyed. In some instances, the participants refer to an absent photograph, such as 'I took a great picture of that, you should see it' which illustrates that experiences are to a great extent visualised, and even an absent photograph may have a certain agency in the situation.

The non-typical: the deviant photograph as a medium for reflection

As mentioned before, interviews with photographs taken by the researcher on the participant's everyday life induce a distance towards everyday routine behaviour that may evoke reflection important for the analysis (Heisley and Levy 1991). Although interactions with participant-produced photographs may not evoke the same type of reflection, since the participant is so familiar with his/her own photographs, reflections upon the act of photography and the choice of motifs occur. Participants may even be surprised by the pictures suddenly appearing during the interview, since they may have forgotten about when and why they were taken. What I found in several of my interviews with backpackers, compared to the photo-shows for an audience, was that the collection of photographs was not always very structured. The interview often started with an apologising comment on how messy the collection is, 'I haven't had the time to order it yet' or 'I have many photos here that are not so interesting for you'. This may be related to how many times the traveller has shown the pictures before and to whom, and how much editing-effort has been invested in the collection. When it comes to the photo-shows in front of an audience, one might assume that the collection is edited according to the self-image that the respondent wants to present, and for this reason it may be edited differently to the one presented to family and friends, or the one presented to a researcher in an interview. Furthermore, one might assume an interplay between the photographic style adopted and the type of technical equipment the tourist has, for instance if s/he has a non-digital camera or a digital camera. For example, with a digital camera it is easier to adopt a snapshot style, and if the tourist is more aesthetically oriented and has the time required for taking specific motifs, s/he may choose another type of camera. These are factors that may affect the level of reflection that has been invested in each photograph before showing them in an interview. However, since the social construction of meaning is situated in a specific spatial and social context, the social construction of tourist experiences in relation to the photographs may take new directions in every situation in which the photographs are displayed. This might be particularly pertinent when it comes to an interview since this is a context that probably deviates most from the 'normal' photo-display situation. Thus, photographs may appear during the interview that the respondent had not paid much attention to previously, and this might be a starting point for reflection that is worth paying attention to as an interviewer.

What emerged in my interviews with backpackers, which was not apparent in the observations, was a new analytical category that became relevant for my study – the non-typical tourist picture. In one of the interviews, the respondent was just about to show me photographs from a train journey, which quite typically included the window view of the passing landscape, as well as his travelling partner in the train. Then suddenly a picture of the train toilet appears, and he laughs and says 'and here is the train toilet'. He reflects upon this slightly odd picture: 'Why did I take a picture of the train toilet, really?' This comment became the starting point for a new analytical theme, since during the course of the project, similar modes of reflecting upon non-typical tourist photos appeared in the interviews. Since these pictures showed situations connected to the travelling situation, such as photographing the interior of a hotel room with laundry hanging in the hotel bathroom, rather than the typical destination pictures, I interpreted these images and their accompanying narrative as 'the travelling scene' as opposed to 'the tourist scene'. Furthermore, images of the travelling scene were an important medium for the construction of an identity as a backpacker, since the harshness and practicalities of travelling is seen as an important experience in itself. Hence, the photo-elicited interview may evoke reflections that an oral-only interview may not, but the non-typical image that suddenly appears in the interview also disrupts the flow of the narrative. This differs from the discursive flow of the typification process described above. Thus, the photograph has a certain agency in this specific situation. It becomes a mediator acting upon the interview-situation as if it has a life of its own, as someone that suddenly interrupts a conversation. Thus, it is not merely a passive recipient of the participant's interpretation of the image. This example shows that how things are said is as important as what is actually said. Furthermore, the how-aspect of narratives and accounts is related to the type of situation where the pictures are shown, that is, the type of social interaction wherein the photographs participate, who participates, for what purpose and what type of editing has been done.

Ambiguity: the gap between the photograph and the ideal image

Photographs are, as several researchers in visual studies have emphasised, ambiguous, implying that the photograph has multiple meanings for different subjects (Pink 2001; Schwartz 1989). Instead of seeing this ambiguity as an obstacle for gaining the real truth 'behind' the photograph (see Pink 2001), ambiguity may be an important starting point for an analysis of the social construction of meaning, particularly when there is an expressed ambivalence in the participant's narrative. Emotional ambivalence may, for example, be evoked by a gap between ideal images and what is portrayed in the photograph. In some studies, especially where the photos are taken by the researcher, the photographic image presented of an interviewee may not be coherent with the self-image s/he prefers to convey. Likewise, in participant-produced interviews with tourists, the constraint embedded in a still photograph on what the traveller actually felt and experienced at the specific moment of photography, sometimes causes frustration and ambivalence. Since tourist photography has a tendency to follow certain socially constructed conventions on what to photograph on what type of occasion (for example, you photograph a nice view and the typical local scene, but to photograph the train toilet is considered deviant), the photographs are sometimes close to the stereotypical. The ambivalence evoked by these might be related to the tension between the traveller's search for individuality and uniqueness, and the norms embedded in a specific tourist culture, such as backpackers (Andersson Cederholm 2004). In a society emphasising individuality, there is often a strong ambivalence connected to social conformity, quite often represented by the typical tourist photographing the typical tourist site. The ambivalence, sometimes taking the form of self-irony, is illustrated in the following quote: 'this is also in Delhi, this is the president's palace. You just had to take a picture of that, but those kinds of pictures are terribly boring! [laughter]'. (ibid.: 232).

Another type of situation that may evoke ambiguity is when the respondent is caught between different photographic styles. Sarah Pink (2001) discusses different photographic categories or styles, and observes that the ethnographer is often in-between, for instance, documentary photography and a leisure-oriented, snap-shot type of photography. Equally interesting though, and often relevant for an analysis of the photographic culture and the life-world of the respondents, is the participants' own photographic styles, and how they categorise and typify themselves. I found, for example, in my study of backpackers, that some respondents were embarrassed about the bad quality of the photographs. 'Normally, I don't take these kinds of photographs' is one type of comment. They were caught between different photographic categories or styles, and had to legitimate their role at the interview situation: 'Here I am a tourist and therefore I take these'.

Social closeness and distance: mediating relationships through the photograph

As mentioned previously, photo-elicited interviews, especially with participant-produced photographs, may ascribe more authority to the participant than traditional oral-only interviews. Furthermore, by the act of self-reflection in the interview, evoked by photographs, the participant invites the researcher to partake in a new direction in the construction of a tourist experience, and thus the social distance between the researcher and the participant diminishes. However, apart from the researcher/participant interaction, which is characterised by more or less closeness or distance, and may consist of a two-part relationship or might include more researchers and respondents, there might be another, third type of actor involved in the conversation. Albeit physically absent, other people that appear on the photographs are often important in the narrative. In narratives of tourist experiences they may be the travel companion or they may be the locals. When it comes to the travel companion, narratives may convey something about their relationship, rather than the actual destination. Since a common way of understanding tourist experiences is often according to categories such as the destination and specific attractions and activities, photographs may reveal those not-so-much-talked about, but often taken for granted experiences such as socialising with your family, partner and friends (Larsen 2006). Photographs may evoke reflections about these taken for granted dimensions of the tourist experience, and they may express social closeness as well as distance.

There is another, not so taken for granted type of relationship but rather regarded as part of the presumed non-ordinary experience of being a tourist: relationships with the locals. Getting close to, or at least interacting with, the locals is an important experience dimension for many tourists, and for the backpackers in my study, close relationship with the locals, preferably over a relatively long time, is an important aspect of an authentic experience. A reverse, but similar type of experience, is illustrated by one of the interviews with the lifestyle entrepreneurs. The owner of a small Bed and Breakfast, while talking about her life and business and relationships with the guests, reaches for a photo album. This album is actually her guest-book, since she found out when she started her business that taking photographs was a convenient way to keep a record. However, the album/guest-book is also a medium for constructing her own tourist-like experiences, since relationships with the guests, or some of the guests, give her experiences of intimacy and authenticity. Thus, she talks about her guests and points at pictures of them, in a manner that is similar to the travellers' narratives on interactions with locals, albeit in this case, the narrator/photographer is the local and the persons portrayed are the tourists (Andersson Cederholm and Hultman 2010). However, in both situations, the relationship between the photographer and the person portrayed illustrates the tension between closeness and distance embodied in the act of collecting experiences of relationships that are often temporary and constrained in time and place (ibid.).

Practical Tips 6.4

ANALYTICAL POTENTIAL OF THE MEDIATING PHOTOGRAPH

The analytical potential of the mediating photograph includes:

- the way the interview participants are doing things with the photographs and the equipment may be as important as what they say;
- narrative typifications as well as the occurrence of non-typical and deviant photography have significance;
- absent photographs may have influence and agency in the interview;
- ambiguity and ambivalence may be analytical points of departure;
- the presence and/or absence of other actors, such as travel companions and locals, may have significance in the construction of experiences.

Ethical and practical challenges

As in all interviews and observations, the use of personal data requires ethical considerations, such as informed consent when personal data is used and careful procedures when coding and writing if anonymity is guaranteed. Researchers often follow these procedures according to general directions from organisations such as the ethical board of research councils. However, the specific social context of data collection always contains challenges and ambiguities. For example, the meaning and interpretation of informed consent and exactly how 'informed' an interviewee usually is and should be about a specific research project is one such controversial issue (Pink 2001). Likewise, the issue of informed consent is often a question of *in situ* interpretation when the researcher is observing semi-public/semi-private events. In some types of events it is both ethically appropriate and practically achievable to inform the participants that they are being observed, but in some, more public-oriented events it is not that self-evident. Furthermore, in line with the common practice of anonymity, publishing photographs of people is a sensitive issue, and many researchers will ask for written permission from the persons appearing in the photograph (ibid.: 43).

Connected to ethical considerations are practical challenges. One such challenge with the photo-elicitation technique is the transcription of oral data connected to specific visual images that is owned and controlled by the participant. Even if the researcher is primarily interested in the narratives, not the visual images per se, the narrative may seem incomplete when the images that the participant refers to are absent. The researcher may then ask for copies of the photographs, which might entail a moral dilemma relating to the intrusion into the participant's personal life and breaking a confidence by asking for copies. However, these copies might be used for coding and

Practical Tips 6.5

VISUALISING ABSENT PHOTOGRAPHS

How can you visualise absent photographs when transcribing oral data?

- If the research participant owns the pictures, ask for permission to copy/photograph the images or film the interview.
- Since publishing photographs entails an ethical dilemma, the copies/film may be used for coding and analysis only.
- translate visual images to oral data during the interview by giving short descriptions of the images.

analysis only, not publishing. Another solution is to photograph or film the photos and/or interview, or to take notes with short descriptions of the images. Being aware of this problem, the interviewer may also give short descriptions of the images during the interview, such as: 'I see here a picture in which you stand in front of a statue and someone is with you . . . Who is . . . ?' Additionally, these types of descriptions may facilitate the interview if they include questions that encourage the interviewee to describe the image in his/her own words.

CHAPTER SUMMARY

- The chapter has presented and discussed photo-elicitation as a method that opens up for analysis the role of the photograph as a social mediator in the interview or observation. With examples from two Swedish studies that focus on the social construction of tourist experiences, the role of photographs was discussed and a practice oriented narrative analysis introduced.
- The notion of authority in the collaborative situation of the interview was discussed by the presentation of two forms of photo-elicitation techniques: interviews with researcher produced and/or introduced photographs and participant-produced photographs.
- The importance of considering the context of production and display was highlighted, such as the degree of formalisation and routinisation of the situation.
- The interactive role of the photograph as an artefact with a certain agency was discussed. How the interview participants do things with photographs and equipment such as computers and projectors, and *how* they talk about them and interact with them, may be as important as *what* they say. Four analytical categories were presented that highlight the interactive context between interviewer, interviewee and photographs: *typifications, the non-typical photograph, ambiguity* and *social closeness and distance*.

- Finally, ethical considerations of interviews and observations with photographs were discussed, and the practical challenges they entail. The problem of transcribing and analysing oral data when the researcher does not have the photographs available were highlighted, and some solutions suggested.

Annotated further reading

Banks, M. (2001) *Visual Methods in Social Research*. London: Sage.
This book also provides a comprehensive introduction to visual ethnographic methods. Although it does not have any lengthy discussions of photo-elicitation, the author points to the role of materiality in visual ethnography, which is important to consider in photo-elicitation.

Holstein, J. A. and Gubrium, J. F. (1997) 'Active Interviewing', in D. Silverman (ed.) *Qualitative Research. Theory, Method and Practice*. London: Sage.
A topic discussed in this chapter, which is related to methodological discussion in general and not specifically connected to photo-elicitation, is the practice-dimension in narrative analysis. Holstein and Gubrium highlight the interactive practices of an interview by emphasising the importance of the interview situation per se as a focus of analysis, not merely as a medium for collecting data.

Oliffe, J. L. and Bottorff, J. L. (2007) 'Further Than the Eye Can See? Photo Elicitation and Research With Men', *Qualitative Health Research*, 17: 850–58.
This article highlights the advantages of photo-elicitation technique when the ambition is to capture emic perspectives, and particularly when the research is about potentially sensitive issues, such as illness experiences. The research presented in the article also points at unexpected results due to the practice-aspect embedded in interviews with photographs.

Pink, S. (2001) *Doing Visual Ethnography: Images, Media and Representation in Research*. London: Sage.
This book provides a comprehensive introduction to visual ethnographic methods. The author's philosophical positioning and critique of a realist tradition in visual ethnography highlights the potential of a constructionist approach in using photographs in interviews.

Schwartz, D. (1989) 'Visual Ethnography: Using Photography in Qualitative Research', *Qualitative Sociology*, 12: 119–54.
Among the publications that focus particularly on photo-elicitation technique, Donna Schwartz's article is an informative and knowledgeable study on the use of photographs in social science. Grounded in her own ethnographic research of a farm community in the US, the text is filled with empirical illustrations and functions well as an introductory text on photo-elicitation technique.

References

Andersson Cederholm, E. (1999) 'The Attraction of the Extraordinary: Images and Experiences of Backpacker Tourists' (In Swedish). Lund: Arkiv Förlag.

Andersson Cederholm, E. (2004) 'The Use of Photo-elicitation in Tourism Research: Framing the Backpacker Experience', *Scandinavian Journal of Hospitality and Tourism*, 4: 225–41.

Andersson Cederholm, E. and Hultman, J. (2010) 'The Value of Intimacy: Negotiating Commercial Relationships in Lifestyle Entrepreneurship', *Scandinavian Journal of Hospitality and Tourism*, 10: 16–32.

Banks, M. (2001) *Visual Methods in Social Research.* London: Sage Publications.

Collier, J. Jr. (1967) *Visual Anthropology: Photography as a Research Method.* New York: Holt, Reinhart and Winston.

Harper, D. (1988) 'Visual Sociology: Expanding Sociological Vision', *The American Sociologist*, 19: 54–70.

Heisley, D. D. and Levy, S. J. (1991) 'Autodriving: A Photoelicitation Technique', *Journal of Consumer Research*, 18: 257–72.

Holstein, J. A. and Gubrium, J. F. (1997) 'Active Interviewing' in D. Silverman (ed.) *Qualitative Research. Theory, Method and Practice*. London: Sage Publications.

Larsen, J. (2006) 'Picturing Bornholm: Producing and Consuming a Tourist Place through Picturing Practices', *Scandinavian Journal of Hospitality and Tourism*, 6: 75–94.

Oliffe, J. L. and Bottorff, J. L. (2007) 'Further Than the Eye Can See? Photo Elicitation and Research With Men', *Qualitative Health Research*, 17: 850–58.

Pink, S. (2001) *Doing Visual Ethnography: Images, Media and Representation in Research*. London: Sage.

Prosser, J. and Schwartz, D. (1998) 'Photographs within the Sociological Research Process' in J. Prosser (ed.) *Image-based Research: A Sourcebook for Qualitative Research*. London: Routledge and Falmer.

Rapley, T. J. (2001) 'The Art(fulness) of Open-ended Interviewing: Some Considerations on Analysing Interviews', *Qualitative Research*, 1: 303–23.

Samuels, J. (2004) 'Breaking the Ethnographer's Frames: Reflections on the Use of Photo Elicitation in Understanding Sri Lankan Monastic Culture', *American Behavioral Scientist*, 47: 1528–50.

Schatzki, T. R., Knorr-Cetina, K. and von Savigny, E. (2001) *The Practice Turn in Contemporary Theory*. London: Routledge.

Schwartz, D. (1989) 'Visual Ethnography: Using Photography in Qualitative Research', *Qualitative Sociology*, 12: 119–54.

Walker, A. L. and Moulton, R. K. (1989) 'Photo Albums: Images of Time and Reflections of Self', *Qualitative Sociology*, 12: 155–82.

7 Video diary methodology and tourist experience research*

Naomi Pocock, Alison McIntosh and Anne Zahra

Introduction

This chapter proposes video diaries as an innovative methodology in tourist experience research. Tourism is a highly visual experience, yet word-based approaches dominate travel and tourism research methods. Video diaries are one innovative methodology to understand the tourist's experience; they capture visual and verbal representation of the participant's reality, as participants choose the scenes worth filming based on personal meaning. Thus, this approach is underpinned by the ontological hermeneutic assumption that understanding and interpretation are part of human existence, and that participants may therefore interpret their *own* experiences. Such participant-driven methods are rare in visual methodologies generally, and even rarer within tourist experience research and, thus, this chapter aims to introduce video diaries as an approach which supplements traditional word-based approaches used in tourism research in order to privilege an insider's perspective of the tourist experience. A case study of research interpreting 'home', a complex, value-laden and personal concept, among returnees from long-term travel, is presented within the chapter to demonstrate the effective application of video diaries and associated issues and challenges.

Background in video methods

Following Bateson and Mead's (1942) pioneering methodology in which they filmed and photographed the Balinese culture over a three year period, scholars have increasingly utilised visual methodologies in scholarly research (Albers and James 1988; Banks 2001; Rakić and Chambers 2010) and ethnographers and anthropologists in particular have used visual methods extensively, from analysis of pictorial media, to ethnographic films scripted by the researcher, to the filming of data gathering, like interviews. Empirically, methods for enquiry have ranged from the scientific capture of natural or experimental situations to the less scientifically controlled, but more genuine/organic participant recording of natural social situations (Asch *et al.* 1973; Gauntlett 1997; Ruby 2000; Sorenson 1976; Worth and Adair 1972). Certainly, this visual approach to social research has now become accepted within these social disciplines (Banks 2001; Collier and Collier 1986; Emmison and Smith 2000; Guindi 2004), though it is still slow to be accepted in sociological research (Chaplin 1994; Holliday 2000); likewise within travel and tourism scholarship.

Indeed, despite the highly visual nature of the tourist experience, purported in particular by Urry's (1990) notion of the 'Tourist Gaze', few travel and tourism research methodologies incorporate visual methods. For example, Jenkins (1999) reviewed different techniques for measuring destination image and lamented the dominance of word-based approaches. Although research methods presented within published research such as the use of diaries and tourist accounts of their own experiences, including memory work, provide depth of insight into tourist's experiences (McIntosh 1999; McIntosh *et al.* 2007; Noy 2004; Small 1999; Tucker 2005), a call for more visual methods is emerging. 'As tourism is, to a large extent, dominated by visual experiences we can regard it as being a rich site for both the 'creation' and analysis of visual evidence' (Feighey 2003: 78). This becomes particularly important as a growing body of research seeks to capture the experiential nature of tourism. However, while some researchers have used photographic methods to understand tourist's experiences (for example, Albers and James 1988; Caton and Santos 2008; Chalfen 1979), video-based methodologies remain relatively scant among published research. This chapter thereby supports the call for more innovative qualitative enquiry around investigations of interaction, human behaviour, culture and, in particular, humanistic and societal dimensions of travel and tourism, by proposing an iterative methodology comprising a series of interviews and a video diary.

Video diaries build on more traditional visual methods like photography, as they capture interaction and the voice of the participant. 'Tourist videos offer an enormously rich vain of visual evidence which could facilitate the kind of "experience-rich tourism research" previously alluded to' (Feighey 2003: 81, spelling in the original). This method allows the author (i.e. the tourist, or in the case study presented here, the 'returnee') to represent their view of reality by letting them decide on scenes worth filming based on their personal significance and meaning. A video diary 'allows respondents more opportunity to represent themselves and provides a greater degree of reflexivity in the research process' (ibid.: 82). Moreover, video diaries are multi-sensory as they capture both the narratives and experiences of the study participants to create comprehensive multifocus research (Collier and Collier 1986; Flick 2006). Thus, video diaries facilitate an in-depth and critical perspective of social phenomena to be captured, by allowing a visual and verbal presence of the respondent without interference from the researcher. Video diaries 'record the facts [and provide] a more comprehensive and holistic presentation of lifestyles and conditions' (Flick 2006: 243). Although the story is inevitably 'skewed by the person of the researcher and their situatedness' (Tribe 2006: 375), a video diarising approach would minimise the influence of the researcher by ensuring the voice of, and aspects of importance to, the individual respondent is retained and reported.

It is proposed that video diaries with supplementary audio explanations as triangulated methods enable respondents to show visually, and explain verbally, the personal meaning and significance of their travel experience. Further to this, follow-up interviews offer them an opportunity to interpret these meanings more deeply. Thus, a holistic, richer, sensory understanding of the tourist's experience can emerge that has not readily been captured in previous research. Therefore, this chapter discusses video diaries as an innovative and complementary method for understanding personal meaning through the tourist experience and draws upon a case study to demonstrate the application of video diaries and associated issues and challenges.

The case study

The case study research aimed to conceptualise 'home' for returnees from long-term travel. The tourism experience has often been conceptualised as anticipation, on-site, and the return home (Botterill and Crompton 1996; Clawson and Knetsch 1966). The concept of 'home' in the twenty-first century is complex, subjective, situated, contextual and value-laden, and is an aspect of the tourist experience not widely examined within tourism research. 'Home' is an important experiential construct within tourism discourse, as tourism and recreation take their very meaning from 'traditional anchors of identity, namely *work* and *home*' (Williams and McIntyre 2001: 392, italics in original) and home and away (White and White 2007). Yet, the concept of home for returnees from long-term travel in particular is uncertain and elusive (Ahmed 1999).

> The decision to return 'home' is fraught with anxiety and uncertainty, due to the implied permanence of the return, the thought of settling down, and the need to make decisions about the future. The return home . . . requires readjustment to life-at-home after 'escaping' for a while through the travelling lifestyle.
>
> (Walter-Pocock 2008: 5)

Therefore, the return 'home' is a distinct experiential phase of travel that can be emotional, traumatic and life-changing. Home, then, is a fundamental conceptualisation requiring deeper understanding within travel and tourism discourse.

Conversational interviews with five returnees from long-term travel to London, France, Germany and Columbia and up to five significant others (friends, parents and siblings) for each returnee, with concentrated active listening on the part of the researcher were used to gather background information, to set the context, to build rapport and to encourage returnees to start conceptualising their 'home'. The initial interview with the returnee was followed by a period of video diarising, whereby the returnee chose the scenes of 'home' to be filmed with little input from the researcher. The respondents were asked to film scenes or talk to the camera as and when they felt appropriate. They were given no directions as to the length of footage, and retained the camera for a period they deemed appropriate (ranging between one and four-and-a-half weeks). Finally, a follow-up interview was conducted with the returnee, whereby a mind map based on the previously collected and summarised visual and audio data was presented by the researcher to the returnee as a framework from which to build a common understanding of the returnee's 'home'.

The pre- and post-video conversational interviews allowed the interviewer to gather rich data and thick descriptions, through the rapport that was inherent in the process (Jennings 2001). Due to the semi-formal format of conversational interviews, deep and detailed data were collected regarding attitudes, opinions and values. Indeed, this methodology is built upon empathic understanding and the researcher built rapport with the participants prior to and during the research process. The kind of information fieldworkers get often depends on the nature of their relationship with participants (Sorenson 1976). Therefore, by clearly explaining the research objective and conceptual importance of the research, using a conversational format, and demonstrating active listening, the researcher was able to build rapport throughout the research process. Such rapport was imperative to enable effective video diarising to take place.

Specifically, in the video diary method employed in this case study, the respondent was loaned a video camera and asked to record visually and verbally scenes that reflect their 'home'. Participants were asked to narrate their video diaries to explain the scenes they were filming. Through the reflection that was inherent in the video prompts, respondents were asked to develop an interpretation of their 'home'. Video prompts included, for example, taking shots of what/who made the returnee feel comfortable or 'at home', things/people that were familiar or made the returnee feel like they wanted to settle here, things that they could identify with, that represented who they were, or that enabled them to express themselves. These things may have included friends, family, mentors, things that represented their culture, a particular geographical location, a political system or organisation, a memory of their old home, something else from their past or something that fitted with their vision for the future.

In contrast, and to gain a more holistic understanding, returnees were also asked to consider what/who made them feel uncomfortable or 'homeless'/not 'at home'; things/people that were unfamiliar or made them feel like they wanted to leave again; and things that they could not understand, that conflicted with who they were, or that prevented them from expressing themselves. Returnees were also asked to compare and contrast here (in this case New Zealand) and overseas whenever it seemed relevant to them.

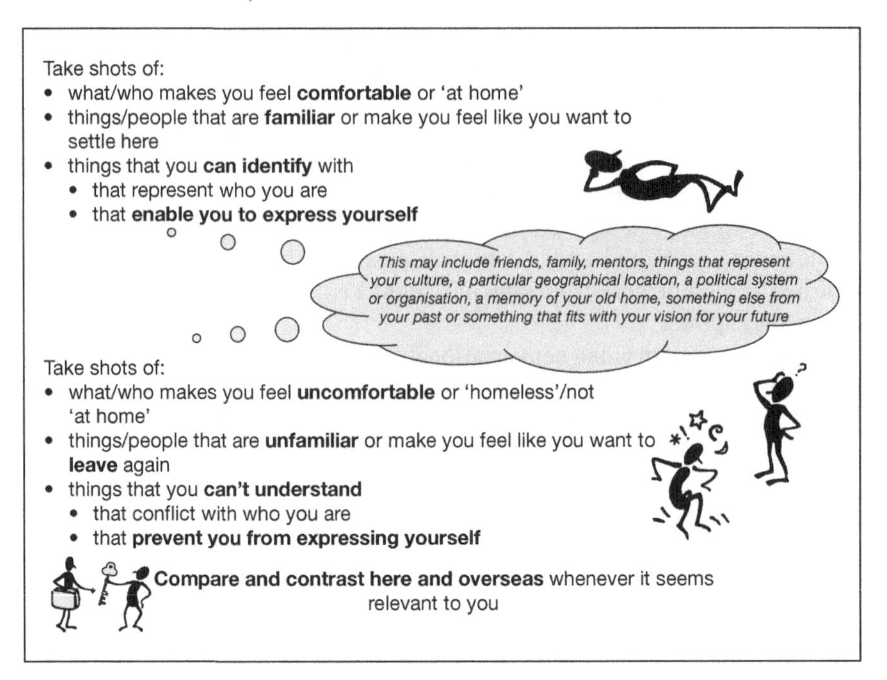

Figure 7.1 Scene prompt flash card

Source: Case study research, used with permission

A flash card was left with the camera to remind participants of these suggestions. This flash card also guided sampling techniques, for example spontaneous versus programmed sampling (Sorenson and Jablonko 1975), discussed later in the chapter, and other filming techniques like talking to the camera, not removing unwanted scenes, asking permission of third parties prior to filming them, and recording scenes from an in-depth and personal perspective. Figures 7.1 and 7.2 exhibit the video prompt flash card (printed double-sided to A6 size and laminated) that was given to respondents in the case study as a guide on what to include and how to film their video diaries.

On the other hand, if returnees were unsure themselves of what 'home' meant to them at that time in their lives, asking them to video their personal reflection of home with a prompt flash card may have been asking too much. Thus, while video diary methods may suit experiential inquiry, the deeper reflection sought in this case study may have been better captured through conversations, with the researcher exploring and clarifying the respondents' thoughts and with the video diary being used as a mechanism to enhance the interview data. Indeed, participants in this case study shared deeply personal experiences in the interviews, such as difficulties reestablishing old friendships, the challenges of deciding what they wanted from their lives and their frustrations with the local culture. Some respondents chose to elaborate on

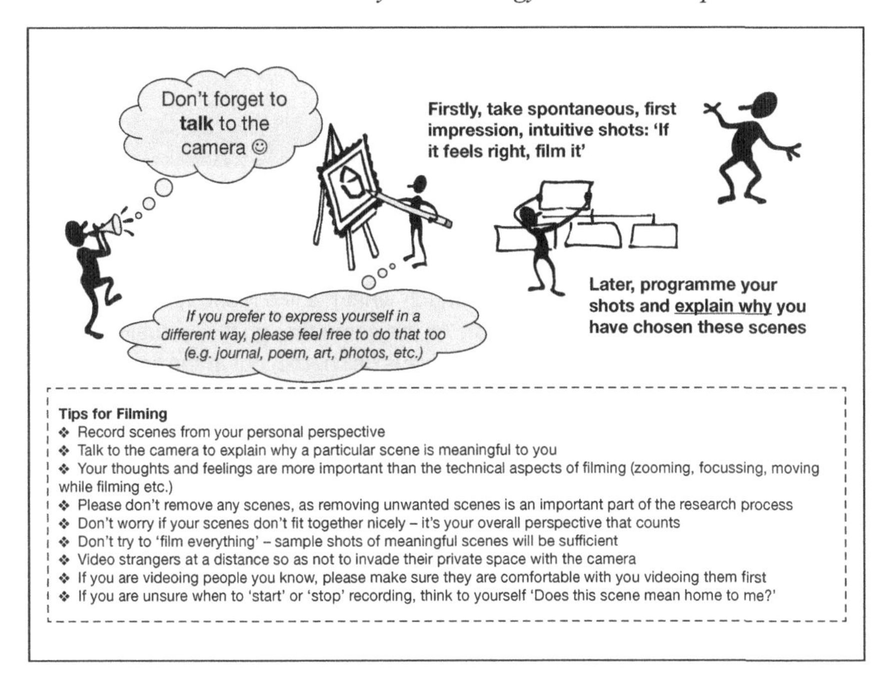

Figure 7.2 Video prompt flash card

Source: Case study research, used with permission

these experiences in their video diaries, for example by filming a place where they grew up while narrating their dislike with what that place represented for them now (see Figure 7.3), or by explaining these deeply personal experiences directly to the camera. Other respondents preferred to film and discuss more generic scenes, for example, their garden, their favourite beach or park, their pet cat, or their sewing studio and chose to explain why they were meaningful to them on their return. Such scenes, while offering a depth of perspective in themselves, did not obviously extend to the deeper personal experiences that were discussed in the interviews and may not have been as meaningful without the interview data to accompany them.

Certainly, context is important with participant-recorded methodology, especially if the researcher is not present while the recording is taking place. The researcher may prompt with questions, to set the context, but the participant works out the interpretation of what they have captured. With this methodology, the participant's voice is dominant, and the researcher's remains a guide (Pink 2001; Ruby 2000). Indeed, this methodology develops knowledge by drawing responses about the social world in a significantly different way to other qualitative methods, by encouraging the respondents to interpret their own social world, by recognising the participant as a co-researcher or equal expert in the field, by the researcher refraining from

Right now you're at Waikanae beach which is an hour north of Wellington and in front of us is Kapiti Island and this is where I spent the first 20 years of my life. As you can see it's a pretty busy day on the beach. There are about five people in the distance. That's pretty much what it's like Monday to Sunday, seven days a week, 365 days a year. All the same though it is actually quite a nice place and it's a place that I will always call home, but at the same time it's a place that I could actually never really live in again. Well, not at this point in my life, I mean it doesn't really fulfill anything that I sort of need or want, but yeah, Kapiti coast, it's a happening place.

Figure 7.3 Beach scene

Source: Research participant's photographs, used with permission

imposing explanations on responses and allowing the researcher only a holistic or overall interpretation of the social phenomenon (Flick 2006). With such methods, the researcher provides an overall analysis and may generate new theory from the research, but the basis of this theory is participant-produced and contextualised (ibid.).

Practical Tips 7.1

RESPONDENT-CREATED VIDEO DIARY

- Give the camera to respondents. Encourage them to record scenes that are meaningful to them. Minimise the researcher's influence on the scene selection.
- Ask participants to narrate their video diaries to explain the scenes they are filming.
- Leave a video prompt card with the camera to guide participants on what to include and how to create their video diary (e.g. combining opportunistic and programmed sampling).
- Remember that for research requiring deeper or personal reflection, a video diary may not suffice and interviews may be required to explore and clarify the respondent's reflections.
- Rapport between the researcher and respondent is imperative.

Thus, the video diary methodology can be considered within, for example, social construction philosophies to understand the participant's representation of their own reality, critical theory to bring about deeper understanding of society, or phenomenology to understand the lived experience and examine the way participants build meaning. In this case study, the choice of method was determined by the need for an emic and holistic understanding of the complex reality of the social phenomenon under study (Walle 1997), that is, the concept of 'home'. The iterative process built a hermeneutic circle of understanding (Caton and Santos 2008; Guignon 2002; Phillips 1996) for both the primary researcher and the returnee, thereby privileging the insider's perspective but generating a deeper and rich, valid and more reliable understanding for the researcher as an outsider. Further to this, it also generated a broader understanding of how the tourism experience fits within the personal meaning and wider life of the individual.

Video diaries as an innovative approach in tourist experience research

Despite the importance of image in the social world (Wagner 2006) and the desired insider's perspective that is manifest in this approach, relatively few researchers hand over the camera to participants, perhaps due to the loss of control and expert status, or perhaps because with this methodology, the 'respondent's account is afforded greater weight in relation to the researcher's than is usual [in academic research]' (Holliday 2000: 519). The first example of researchers relinquishing the control of visualisation entirely to participants is Worth and Adair's (1972) Navajo study. This study offered revolutionary development of emic understanding, for example, Worth and Adair realised the importance of walking to the Navajo people, which initially seemed like 'wrong filmmaking' (Worth and Adair 1972: 144) but later emerged as a significant aspect of their culture, 'an important event in and of itself and not just a way of getting somewhere' (ibid.: 146). Similarly, Carelli (1998) has facilitated Brazilian Indians use of videomaking to foster 'indigenous cultural and political self-awareness and autonomy' (Aufderheide 1995: 83) as participants experienced a revival of tribal rituals. Such ethnographic methods, despite amateur filming techniques and far from patronising primitivism or contributing to cultural decay as some critiques would argue, in fact act as a form of advocacy by placing the culture 'in the context of a struggle for political, economic and cultural space' (ibid.: 87). Such participant-driven elicitation approaches in tourism may also offer significant insights into the minds of both tourists and their hosts, and indeed into their impact on society as well as the impact of travel on their own life. Certainly, in the present case study, the video diaries provided a mechanism through which the primary researcher could capture and thus share her in-depth understanding of the participants she interviewed.

Such an approach corresponds to the hermeneutic philosophy of allowing the participant to interpret their *own* experiences. Indeed, hermeneutic inquirers establish a context and meaning for people's experiences; they construct the 'reality' based on their interpretations of data provided by participants of their perceptions of experiences (Patton 2002). Specifically, hermeneutics questions the cultural, temporal and spatial contexts under which a human act takes place that make it possible to interpret its meanings. Thus, hermeneutics appears to be an appropriate paradigm for tourist-experience research as it incorporates the aspects of constructivism and phenomenology. Moreover, hermeneutics recognises, and in fact requires reflection on, the influence of the researcher in the research process. All researchers stem from a particular perspective, standpoint or situational context. The extent to which these influence their interpretations may differ among paradigms, depending on how far they attempt to distance themselves from the research, but untrustworthy or invalid results emerge if a researcher does not even attempt to acknowledge this influence. Considering hermeneutic ontology and interpretive epistemology, a qualitative research method in the form of video diaries and conversational interviews may thus enable the researcher to gain a *reflective* insider's perspective. Together, visual along with audio or textual methods provide a more powerful means of interpretation than one or other of these methods can provide alone. While visual records improve the depth of understanding, verbal or textual descriptions are needed to make sense of visual records.

Commonly, visual methods alone are not enough as they do not reproduce reality (Flick 2006). Due to the individual's forestructure and the structuralist nature of social interpretation, scenes may be interpreted in different ways requiring contextual explanation by the author. Moreover, by capturing a visual record of the scene, reality is transferred into a frame, thus potentially distorting reality. For example, in their reflection of 'home' in the case study research, respondents may film the main street of their local town. Certainly, in seeing the main street for him/herself in the form of video footage, the researcher is able to clearly envisage and therefore understand this view of the respondent's description of home. However, while one respondent may perceive this main street as dead, slow, boring or frustrating, due to the relative lack of people after the London crowds, for example, another may describe this view of home as relaxing, peaceful and quiet after the same London experience (Walter 2006). Alternatively, perhaps the people on the street are simply outside the frame of the video camera. Therefore, the subjective influences on the selective material must be documented, for example by encouraging respondents to provide narratives to describe scenes they capture through video to set the context and provide interpretation in relation to their personal meaning. Indeed, the respondent's identity, objectives and desires are an integral part of any hermeneutic mode of inquiry, and by allowing the informant to conduct the filming in this research, the desired insider's perspective will be obtained. Verbal or textual explanations of the scene from the respondents and subsequent interpretation by both the respondent and researcher are vital to accompany the visual image, to ensure accurate and trustworthy interpretation of social phenomena.

Other studies have successfully used video technology alone to meet research objectives, for example, Nicholson (2006) who analysed silent amateur home movie footage from tourists in the 1930s Mediterranean to understand how twentieth-century touristic visual practises have evolved; Heath and Hindmarsh (2002) who used a stationary camera to record medical consultations, this approach allowing them to leave the setting and ensure the participants were less distracted; Rakić and Karagiannakis (2006) who made an academic ethnographic documentary by filming their fieldwork (interviews and participant observation) to record the impact of tourism on Crete; Schaeffer (1975), who videoed the effects of marijuana use on behaviour in a remote Jamaican village; and Threadgold (2000) who used video diaries and interviews in a longitudinal study on aging Vietnamese women's self-representation. Others have examined the effect of visual media on tourist behaviour (for example, Dunn 2006; Iwashita 2006). However, providing respondents with the opportunity to record their own scenes, as suggested in this chapter, is integral to the concept of *Dasein* (the concept that understanding and interpretation are fundamental features of man's Being) in hermeneutic philosophy (Gadamer 1975; Guignon 2002; Phillips 1996).

Critics of video methods may condemn the subjective nature of film, as the choice of scenes depends on the film-maker's personality, selectivity, objectives and desires and may therefore be deemed unscientific (e.g. Byers 1966 cited in Sorenson 1976). Censorship may also occur in terms of the author leaving out significant information, either deliberately or through poor filming techniques, like inappropriately deciding when to start and stop recording, moving while recording, or selecting an inappropriate focal length or optical focus (Flick 2006). However, in any research investigation human selection plays a decisive role. Indeed, 'to film at all is to film selectively' (Sorenson 1976: 254). As credibility and authenticity are key with social research, 'some degree of control . . . by a person behind the camera is essential if rich visual data are to result' (Collier and Collier 1986: 148). Moreover, artistic films are not necessary for ethnographic film to be effective, indeed the character of an anthropological film may be lost if the camera is operated by a professional film-maker. 'We do not demand that a field ethnologist write with the skill of a novelist or a poet . . . it is equally inappropriate to demand that filmed behaviour have the earmarks of a work of art' (Mead 1975: 5). Indeed,

> the white middle-class Western eye, conditioned by its culture and the intricate technology and tradition of Hollywood and the television screen, is in danger of losing sight of the beauty and vitality of the film produced simply and under the control of the filmmaker for personal expression.
>
> (Worth and Adair 1972: 261)

Therefore, the prompt for video diaries would remind respondents that the importance of deep personal reflection overrides the technical aspects of filming (see Figure 7.2).

Moreover, rather than trying to 'film everything', respondents may be asked to use a combination of opportunistic and programmed samples to record their experience (Sorenson and Jablonko 1975). First, respondents may be encouraged to take spontaneous, first impression and intuitive shots based on partially formatted ideas of 'what feels right'. This approach documents naturally occurring phenomena without the structured mind of the film-maker influencing the film. Respondents may also be asked to undertake programmed filming, whereby the author decides where, what and when to film, and explains why he/she has chosen those locations, times and scenes. The nature of the perspective with programmed sampling is less important than the explanation of the scene. This approach breaks the egocentricity of opportunistic sampling, as it is more ethnocentric and based on culture, background, learnt concepts, ideas and values (ibid.). This approach is limited to preconceived ideas of what is important, which in itself provides insight into the respondent's perspective. Therefore, by encouraging both opportunistic and programmed sampling, it is acknowledged that each method skews the sample in a different way, but that together they balance and increase the informative potential of visual records. On the other hand, respondents may choose to ignore such 'guidance' and film how and what they feel like, resulting in a rich array of diverse video diaries across the sample, as sometimes occurred in the case study presented in this chapter.

Indeed, as video diaries capture the emic, a diversity of content, structure and depth is likely to emerge among respondents as the ability to capture the personal depth and reflection may also depend on the respondent's personality, the length of time they have with the camera, their technical ability, and degree of confidence with a camera. For example, in the case study research, some respondents needed more time than others (from one to four-and-a-half weeks) with the camera to reflect on their concept of 'home' or to film their 'home' within a logistically busy lifestyle. Other respondents were not naturally or openly reflexive people, and thus their reflections of 'home' emerged in more detail in the interviews with prompting and clarifying by the researcher than in the participant-driven video diaries. Some respondents preferred to talk directly to the camera, and in one instance to read aloud to the camera from a personal diary, while others preferred to remain behind the camera, filming scenes that reflected their concept of home, including a-day-in-the-life-of type footage. Further research is needed in this area to explore respondent's characteristics and their impact on the quality and reliability of video diary data. Certainly, innovative research methods used in the constructivist and interpretive paradigms need to consider the personal nature of video diaries. Of course, such diversity reflects the qualitative richness of social research and may be viewed as a strength rather than a limitation.

Moreover, the researcher's observations, research questions and questioning techniques may be more important than the quality or content of the film produced. Video may sharpen our vision, but 'insight is a product of acuity of human perception' (Collier and Collier 1986: 207). The video diary adds to the insight, but the interpretation cannot be made on the diary alone. Analysis is an important part of the research process, and video diaries with accompanying interviews may be analysed using a vast array of approaches. For example a positivist approach may count the number of times a word, like 'home', appears (or indeed is missing) from the video and interview transcripts. A narrative analysis might explore the story-telling, confessional nature of the video diary (in the present study, one respondent talked for 1 hour and 25 minutes in one uninterrupted scene into the camera) and interviews. A metaphorical analysis might examine the meaning of shorter films of, for example, hobbies, people, places and pets in the social constructions of 'home'.

Furthermore, in terms of video analysis and the subsequent presentation of video footage, no one technique dominates and indeed there is 'no method of interpretation for [video] material which deals directly with the visual level' (Flick 2006: 241) as films are understood as visual texts and analysed as such (ibid.). Moreover, the traditional presentation of academic material in printed formats precludes more interactive and 'insider' interpretations via multimedia formats, such as blogs, documentaries and social networking sites. Thus, in the current academic environment, the presentation of video diaries may be limited to such approaches as film strips (see Figure 7.3) with accompanying transcribed text. However, these approaches lose the depth and richness of the video diary that the methodology originally sought. For example, in Figure 7.3 the respondent's sarcasm is diminished in the filmstrip format as the original footage more clearly illustrated the deserted atmosphere in the pan of the beach scene he was discussing and the sarcasm could be heard in his tone as he narrated the scene.

Thus, theoretically, with an initial interview, the circle of hermeneutic interpretation commences with the respondent doing the talking to avoid researcher-led discussions. In the initial interview, the researcher also steers the respondent through questioning towards a conceptualisation of their experience, and thereby enables them to start the next phase: the video diary. As tourist experiences are complex, subjective and value-laden constructs, the benefit of the video diarising aspect of this methodology is generating understanding by offering the participant the opportunity to represent their own reality through visual methods, providing the outsider-researcher with context to accompany interview conversations, and privileging the insider's perspective, which is a fundamental component of hermeneutical *Dasein* and integral to qualitative research philosophy. With the follow-up interview, the hermeneutic circle of interpretation is complete as the researcher and participants develop together an interpretation of the experience, based on the video diary and previous interviews.

Practical Tips 7.2

INTERPRETING VIDEO DIARIES

- Remember that one scene can be interpreted in different ways by different people; therefore, be sure to ask the respondents for *their* interpretation of a scene.
- Be open and flexible to ethnographic (and thus original, unique) footage – do not expect an artistic or 'Hollywood' creation.
- Be open to video diaries (the products) varying greatly among respondents.
- Remind respondents that the importance of deep personal reflection overrides the technical aspects of filming.

Ethics and limitations

The video diary method on its own may be insufficient to explore a deeply personal and reflective research objective if the concept under reflection is itself elusive and uncertain, such as the concept of 'home' for returnees from long-term travel. Indeed, the depth of the personal experience may be better captured through other methods, like in-depth conversational interviews. Moreover, video diaries may be limited by the inability to capture personal depth and reflection due to the respondent's personality, the length of time respondents have with the camera, their technical ability and degree of confidence with a camera. The main contribution of the video diaries may thus comprise the voice commentaries and reflections that accompany the visual images. Therefore, video diaries may supplement other methods, for example written diaries, journalling, photographic material, poetry, or participant observation, rather than the video diary providing the depth of data sought in emic research, depending on the research objective. Furthermore, to achieve the depth of personal experience sought with emic research, rapport between the researcher and respondent is imperative. Empathic understanding as a basis of rapport is primarily achieved through active listening; therefore, an opportunity for the researcher to actively listen to the respondent and thereby build rapport may be necessary before the depth of personal stories can emerge within video diaries.

Furthermore, the sharing of such deeply personal experiences, whether through interview conversations or video diarising, inevitably raises ethical considerations in terms of the researcher's responsibility towards respondents. Revealing in-depth personal reflections may be cathartic for participants (Zahra and McIntosh 2007) as the researcher actively listens to the respondent's emotional deluge; however, the consequent possible impact on the emotional needs of the researcher herself may remain unresolved. For example, consideration may need to be given to how the researcher will be supported, whether through a supervisory panel, a peer support group or even through counselling, as she attempts to understand the deeply personal journey of the participants.

Moreover, in the case study presented in this chapter, some respondents disclosed negative emotional aspects of 'home' with which the returnee was still coming to terms, and for which they perhaps needed professional support. Returnees also filmed family members or friends and later discussed those relationships with the camera and/or the researcher, thus presenting a selective or biased view of the relationship that did not incorporate the other's perspective. These findings raise the question of the ethical responsibilities of researchers in experiential research. Nevertheless, if all respondents participate voluntarily, understand the personal nature of the research questions prior to commencing the research, and may withdraw at any stage up to data analysis, some ethical considerations may be adequately addressed from the respondent's perspective at least.

CHAPTER SUMMARY

- Video diarising minimises the influence of the researcher and privileges the situation of the individual respondents to capture an insider's perspective. Video diaries may produce best results when accompanied by other methods such as interviews.
- Video diaries inevitably raise ethical concerns, such as the responsibilities of researchers in experiential research, that need to be addressed as part of the research methodology.
- Despite the highly visual nature of the tourist experience, very little travel and tourism research incorporates visual methodology and scholars are encouraged to explore innovative visual technologies. Some researchers relied on visual methods to produce ethnographic documentaries and disseminate their research findings to wider audiences (see discussions in Rakić and Chambers 2010). These were not one of the main aims of this chapter's case study, which aimed to generate 'new knowledge' through a deeper understanding of returnees from long-term travel and their concept of 'home'.
- Flick (2006) argued that interview conversations flow more naturally in settings with less obtrusive technology intrusion. This being the case, dictaphones can be used in interviews and video diaries can be created separately, similar to the way in which these were used in the case study included in this chapter. Visual methods should aim to fulfil an epistemological objective and recording technology should be restricted 'to the collection of data necessary to the research question and the theoretical framework' (ibid.: 284).
- The sociology literature portrays visual data collection and analysis as complementary to other data sources and analyses (Chaplin 1994; Pink 2001) as was the case in the case study discussed in this chapter. More research is required to establish whether video diaries can be used on their own to meet deeply personal and reflective research objectives.

- The visual is important in the acquisition of human knowledge in everyday circumstances; yet, in an academic context collecting, analysing and disseminating the visual can be limited and problematic. Perhaps the critical question needs to be raised as to whether academic traditions are in concordance with the social phenomena scholars are trying to understand.

Annotated further reading

Collier, J. and Collier, M. (1986) *Visual Anthropology: Photography as a Research Method*. Albuquerque: The University of New Mexico Press.
This book contains some useful chapters on film and video, including theoretical arguments for film in research, considerations when entering the field, and a series of chapters regarding the analysis of film footage.

Feighey, W. (2003) 'Negative image? Developing the visual in tourism research', *Current Issues in Tourism*, 6: 76–85.
This is a good introductory article for those interested in tourism-related visual methodologies. Feighey draws on western concepts of 'looking', 'seeing' and 'knowing' to argue for more visual methodologies in tourism research, and specifically calls for more tourist-created and researcher-created video evidence, including video diaries.

Flick, U. (2006) *An Introduction to Qualitative Research*. London: Sage.
This is a good introductory book on qualitative methods and contains a useful chapter on visual methods in which Flick briefly describes in general terms some problems, contributions and limitations of video methods.

Hockings, P. (ed.) (1975) *Principles of Visual Anthropology*. Mouton Publishers, Paris.
This is an excellent book for beginners to visual methods, and includes philosophies, theories and practical examples of visual methodologies in anthropology. The reader is directed in particular towards Mead's introductory chapter and Sorenson and Hocking's concluding chapters.

Noyes, A. (2004) 'Video diary: a method for exploring learning dispositions', *Cambridge Journal of Education*, 34: 193–209.
This is a useful article elaborating on the theories, practicalities and data analysis of the video diary method, specifically in the context of children's learning environments.

Worth, S. and Adair, J. (1972) *Through Navajo Eyes: An Exploration in Film Communication and Anthropology*. Ontario: Indiana University Press.
This was one of the original attempts at participant-driven motion picture methods. Worth and Adair's insightful comments on their assumptions of 'wrong film-making' in the narrative style chapter are particularly enlightening.

Note

* Sections of this chapter were originally published in Pocock, N., Zahra, A., and McIntosh, A. (2009) 'Proposing video diaries as an innovative methodology in tourist experience research', *Tourism and Hospitality Planning and Development*, 6: 109–119 and have been reproduced here with permission from the publishers.

References

Ahmed, S. (1999) 'Home and away: narratives of migration and estrangement', *International Journal of Cultural Studies*, 2: 329–47.

Albers, P. C. and James, W. R. (1988) 'Travel photography: a methodological approach', *Annals of Tourism Research*, 15: 134–58.

Asch, T., Marshall, J. and Spier, P. (1973) 'Ethnographic film: structure and function', *Annual Review of Anthropology*, 2: 179–87.

Aufderheide, P. (1995) 'The video in the villages project: videomaking with and by Brazilian Indians', *Visual Anthropology Review*, 11: 83–93.

Banks, M. (2001) *Visual Methods in Social Research*. London: Sage.

Bateson, G. and Mead, M. (1942) *Balinese Character: A Photographic Analysis*. New York: New York Academy of Sciences.

Botterill, T. D. and Crompton, J. L. 1996, 'Two case studies exploring the nature of the tourist's experience', *Journal of Leisure Research*, 28: 57–82.

Brownell, J. (1986) *Building on Active Listening Skills*. New Jersey: Prentice-Hall.

Byers, P. (1966) 'Cameras don't take pictures', *Columbia Univ. Forum*, 9(1): 27–32.

Carelli, V. (1998) 'Video in the villages: utilization of video tapes as an instrument of ethnic affirmation among Brazilian Indian groups', *Commission of Visual Anthropology Newsletter*, May: 10–15.

Caton, K. and Santos, C. A. (2008) 'Closing the hermeneutic circle? Photographic encounters with the other', *Annals of Tourism Research*, 35: 7–26.

Chalfen, R. 1979, 'Photography's role in tourism', *Annals of Tourism Research*, 6: 435–47.

Chaplin, E. (1994) *Sociology and Visual Representation*. London: Routledge.

Clawson, M. and Knetsch, J. L. (1966) *Economics of Outdoor Recreation*. Baltimore: The John Hopkins Press.

Collier, J. and Collier, M. (1986) *Visual Anthropology: Photography as a Research Method*. Albuquerque: The University of New Mexico Press.

Dunn, D. (2006) 'Singular encounters: mediating the tourist destination in British television holiday programmes', *Tourist Studies*, 6: 37–58.

Emmison, M. and Smith, P. (2000) *Researching the Visual*. London: Sage.

Feighey, W. (2003) 'Negative image? Developing the visual in tourism research', *Current Issues in Tourism*, 6: 76–85.

Flick, U. (2006) *An Introduction to Qualitative Research*. London: Sage.

Gadamer, H. G. (1975) *Truth and Method*. London: Sheed and Ward.

Gauntlett, D. (1997) *Video Critical: Children, the Environment and Media Power*. Luton: John Libbey Media.

Guignon, C. (2002) 'Truth in interpretation: a hermeneutic approach', in M. Krausz (ed.), *Is There a Single Right Interpretation?* Pennsylvania: The Pennsylvania State University Press.

Guindi, F. E. (2004) *Visual Anthropology: Essential Method and Theory*. California: AltaMira Press.

Heath, C. and Hindmarsh, J. (2002) 'Analysing interaction: video, ethnography and situated conduct', in T. May (ed.), *Qualitative Research in Action*. London: Sage.

Holliday, R. (2000) 'We've been framed: visualising methodology', *The Sociological Review*, 48: 503–21.

Iwashita, C. (2006) 'Media representation of the UK as a destination for Japanese tourists: popular culture and tourism', *Tourist Studies*, 6: 59–77.

Jenkins, O. 1999, 'Understanding and measuring tourist destination images', *International Journal of Tourism Research*, 1: 1–15.

Jennings, G. (2001) *Tourism Research*. Milton: Wiley.

McIntosh, A. (1999) 'Into the tourist's mind: understanding the value of the heritage experience', *Journal of Travel and Tourism Marketing*, 8: 41–64.

McIntosh, A., Harris, C. and Walter, N. (2007) 'Traveller repatriation: a dialectical contribution to tourist experience research', *Critical Tourism Studies Conference*, Split, Croatia.

Mead, M. (1975) 'Visual anthropology in a discipline of words', in P. Hockings (ed.), *Principles of Visual Anthropology*. Paris: Mouton Publishers.

Nicholson, H. N. (2006) 'Through the Balkan States: home movies as travel texts and tourism histories in the Mediterranean', *Tourist Studies*, 6: 13–36.

Noy, C. (2004) 'This trip really changed me: backpackers' narratives of self-change', *Annals of Tourism Research*, 31: 78–102.

Patton, M. Q. (2002) *Qualitative Research and Evaluation Methods*. London: Sage Publications.

Phillips, J. (1996) 'Key concepts: hermeneutics', *Philosophy, Psychiatry, and Psychology*, 3: 61–9.

Pink, S. (2001) 'More visualising, more methodologies: on video, reflexivity and qualitative research', *The Sociological Review*, 49: 586–99.

Rakić, T. and Chambers, D. (2010) 'Innovative techniques in tourism research: an exploration of visual methods and academic filmmaking', *International Journal of Tourism Research*, 12: 379–89.

Rakić, T. and Karagiannakis, Y. (2006) *Of Holidays and Olives* [a 38-minute ethnographic documentary]. UK and Greece: PitchDarkProductions.

Ruby, J. (2000) *Picturing Culture: Explorations of Film and Anthropology*. Chicago: University of Chicago Press.

Schaeffer, J. H. (1975) 'Videotape: new techniques of observation and analysis in anthropology ', in P. Hockings (ed.), *Principles of Visual Anthropology*. Paris: Mouton Publishers.

Small, J. (1999) 'Memory-work: a method for researching women's tourist experiences', *Tourism Management*, 20: 25–35.

Sorenson, E. R. (1976) *The Edge of the Forest: Land, Childhood and Change in New Guinea*. Smithsonian Institution Press: Washington D.C.

Sorenson, E. R. and Jablonko, A. (1975) 'Research filming of naturally occurring phenomena: basic strategies', in P. Hockings (ed.), *Principles of Visual Anthropology*. Paris: Mouton Publishers.

Threadgold, T. (2000) 'When home is always a foreign place: diaspora, dialogue, translations', *Communal/Plural*, 8: 193–217.

Tribe, J. (2006) 'The truth about tourism', *Annals of Tourism Research*, 33: 360–81.

Tucker, H. (2005) 'Narratives of place and self: differing experiences of package coach tours in New Zealand', *Tourist Studies*, 5: 267–82.

Urry, J. (1990) *The Tourist Gaze: Leisure and Travel in Contemporary Societies*. London: Sage.

Wagner, J. (2006) 'Visible materials, visualised theory and images of social research', *Visual Studies*, 21: 55–69.

Walle, A. H. (1997) 'Quantitative versus qualitative tourism research', *Annals of Tourism Research*, 24: 524–36.

Walter-Pocock, N. (2008) 'Roof, relationships, roots: a hermeneutical understanding of why returnees from long-term travel may be considered homeless', *Council for Australian University Tourism and Hospitality Education* (CAUTHE).

Walter, N. (2006) *An Investigation into Travellers Repatriating to New Zealand, having completed their OE*. Masters Dissertation Thesis, University of Waikato.

White, N. R. and White, P. B. (2007) 'Home and away: tourists in a connected world', *Annals of Tourism Research*, 34: 88–104.

Whorf, B. (1926) *Language, Thought and Reality*. Cambridge: MIT Press.

Williams, D. R. and McIntyre, N. (2001) 'Where heart and home reside: changing constructions of place and identity', in *Trends 2000: Shaping the Future*, Lansing, MI: Michigan State University, Dept of Park, Recreation and Tourism Resources, pp. 392–403.

Worth, S. and Adair, J. (1972) *Through Navajo Eyes: An Exploration in Film Communication and Anthropology*. Ontario: Indiana University Press.

Zahra, A. and McIntosh, A. (2007) 'Volunteer tourism: evidence of cathartic tourist experiences', *Tourism Recreation Research*, 32: 115–19.

8 The drawing methodology in tourism research

William Cannon Hunter

Introduction

This chapter introduces the basic theory and practice of the drawing methodology in tourism research. This approach to visual research focuses on understanding peoples' held subjectivities toward, or impressions and perceptions of persons, places or things encountered or imagined in tourism. The state of the art of visual methods and the drawing methodology in tourism research is briefly introduced before moving on to a discussion regarding the logic of drawing in anthropology, the arts and the social sciences. The contours of the practice of drawing methodology-based research in anthropology, sociology, psychology and tourism research are then described. An example of the methodology follows, based on a study of an iconic cultural artefact of Jeju Island, Korea known as the "stone grandfather" or "dolhareubang" which is one of the most commonly recognized representations of the island's traditional culture. Finally, an in-depth overview of the theory and practice of the drawing methodology is discussed, with a description of the critical issues, or pitfalls. The chapter concludes with a summary that highlights several key points that the reader should take away from the chapter. This chapter, "The drawing methodology in tourism research", highlights by means of example the importance of the theoretical and interdisciplinary underpinnings of visual methods in tourism studies. Visual methods are not only another way to do research in tourism; they represent a new paradigm or way of thinking more deeply about the social and cultural effects of tourism.

Visual methods and the drawing methodology

The drawing methodology is a form of visual research used to understand peoples' impressions, perceptions or held subjectivities toward persons, places or things. As a research methodology, drawing links linguistic and pre-linguistic meanings together to get to the root of perceptions concerning the personal, the social or the cultural and in addition, links so-called qualitative and quantitative approaches to social science. Drawing, along with other visual methodologies, is a relatively unexplored research method in tourism research although in the fields of anthropology, psychology and the arts it has been in common practice (in one form or another) for decades.

Books on visual research published in the last decade have focused on visual culture and representation (Crouch and Lübbren 2006; Dikovitskaya 2006; Sturken and Cartwright 2004; Hall 2002) by emphasizing the cultural significance of the photograph in popular media. Otherwise, the focus has been on the mechanics of using video and photography in research (Pink 2007; Banks 2001; Collier and Collier 1999) or strictly on the analysis of photographic images (Rose 2007; van Leeuwen and Lewitt 2001). The visual was tacitly relegated to the field of media studies where semiotic analysis of existing images or the technical elements of photography and cinematography are the topics of interest. The first-hand creation of images, in the form of sketches, portraits, landscapes, maps and diagrams is seen, possibly, as the exclusive domain of the expert artist or draftsperson, or conversely, as belonging to the realm of the child or the pathologically impaired. Otherwise when drawing is mentioned (Lawrence-Lightfoot and Davis 1997), it is as a metaphor for a philosophy of qualitative research. In any sense, at first glance it would not seem to be an enterprise of much value to tourism research.

The drawing methodology has probably been avoided in tourism research because of certain ontological or epistemological problems that divide so-called conventional socio-scientific research from "alternative" approaches (Russell and Ison 2000; Checkland 1985). On the one hand, there is the notion that reality is objective and concrete, and on the other hand there is a notion that social realities are multiple, relative and subjective even though the operational and philosophical relationship between these two perspectives has long been more or less resolved (Berger and Luckman, 1966).

Drawing is actually about the simultaneous and self-reflexive expression of an individual's subjectivity (skills, knowledge, memory, perceptions and attitudes) and the objectivity of the drawing created (its representational verisimilitude or visual correspondence to perceived experience). In a drawing, objective reality and subjective perceptions are joined together in a way that allows researchers to understand "constructions of social reality" rather than "the reality of social constructions" (Hall 2002). It is an exciting research approach that is more quantum than Newtonian in design and allows researchers to get at the social relativity of tourism as an equally global and local phenomenon.

The logic of drawing

The act of drawing is the formation of non-linguistic signs or images that work as communicative devices in conjunction with verbal cues or anchors. They can logically and legibly depict visual events but vary in meaning depending on the deliberate or subliminal intentions of the author. Cultural values and norms, ritual, social power, personal histories and experiences and various hidden or latent elements of the imagination combine with attempts at "representational verisimilitude" in a highly personal and unique expression. In other words, drawings are representations and originate as "things-in-themselves" (of seen, imagined or remembered people, costumes, food, features of the natural and built environment and surroundings or other unique iconic features) before they are transformed as drawn expressions. Drawings should be "true" descriptions (Brown 1995; Wolcott 1995) (the drawing methodology is based on an assumption that participants will attempt to draw the subject of research "as it appears in reality"), but may express at least as much as their creators' conscious or unconscious subjectivities as they do about their subject and for this reason a drawing requires verbal anchors or a shared context to avoid enigmatic or deceptive readings by others.

Consider the 19,000-year-old cave paintings at Chauvet or the 32,000-year-old depictions at Lascaux, in France (Curtis 2006; Guthrie 2005). The handprints, paintings of animal and human figures, and abstract geometric symbols are legible, inspiring, beautiful, but their creators' ritualistic or symbolic intentions are long lost. When Picasso saw them he reportedly said to his guide, "They've invented everything" (Thurman 2009), but without knowing the conventional relationship between symbol and its context, these cave paintings are only "pictures without meaning" and their interpretation is problematic and controversial if not impossible.

Likewise, Greek realism, Byzantine iconography, abstract impressionism or even computer simulations no matter how "real" they seem to the period cannot be reliably interpreted without knowledge of linguistic and socio-historical context. Consider how indecipherable the work of Kandinsky (1866–1944), or Picasso (1881–1973) would be without an understanding of their intentions. For Kandinsky, the internal, subtler emotions a painting evoked were more important than its representational qualities (Kandinsky 1977), and for Picasso it was the surprise, the accident in the act of painting that was valued (Ashton 1972), not representation. A painting (or drawing) can convey the artist's experience while simultaneously evoking the unique and diverse subjectivities of multiple viewers. But for a drawing to have any larger shared meaning it must be accompanied by an authentic verbal interpretation.

In social scientific research, drawings correspond to relativistic and dynamic (Gallarza *et al.* 2002) networks of meaning. Their representational integrity is contingent on the skills and social conditioning of their author and can be abstract, childlike or highly sophisticated maps, diagrams or even computer generated simulations. Understanding the latent subjectivity in a drawing, however, requires a collaborative effort between "artist" and "interpreter" to identify if the drawing of a brown horse, for example reveals that, "I think I see a brown horse" or that "I do not like brown horses".

In a semiotic sense (Metro-Roland 2009; Peirce 1931), drawings are like words in that they are signs that have connotative or denotative values. In tourism the word "tourist" denotes a visitor or a traveler as defined by the

UNWTO or other authority. But "tourist" also has many connotations (certain emotional and imaginative associations)—irritating, domineering, loud, rich, stupid, interested—depending upon the conditions, or context, in which such meanings are exchanged. Likewise in a drawing of a tourist, certain denotative elements such as gender, attire, behaviour and relationships and other pictorial elements would combine with the artist's compositional devices such as size, perspective and proportion that could connote certain social power relations such as importance, acceptability and even fear or dislike. And it is in this sense that the drawing methodology has merit as a means to understanding the value-judgments of respondents.

Drawing in the social sciences

In the social sciences drawing has been employed in anthropology for the study of the so-called "non-European races" (Paget 1932), and in psychology for the study of the pathological (Machover 1949). In both traditions it has been assumed that the subject's art (children, the pathological or the "tribal") is inherently flawed (Perez 1988), and that art activities such as drawing are not as important as the things they reveal about a person (Leeds 1989: 102). Currently this aspect of the tradition still holds true. The things that drawing reveals about the respondent are more highly valued than the virtuosity of the drawing itself.

In the past, research has focused on identifying the underlying or latent pathologies in research subjects who are "primitive" ("non-European" or developing populations), "dispossessed" (women and children) or "deviant" (prisoners). In anthropology the focus has been on demonstrating the cultural handicap of "primitivism" (Martlew and Connolly 1996; Mead 1953). In analytical psychology the focus has been on the distressed, especially women (Jung 1968), and in cognitive psychology the focus has been on the pathological, especially in terms of cognitive development or deviance.

In Machover's Draw-a-Person test (Machover 1949), respondents are asked to "draw a person" and then inferences are made based on characteristics of the drawing. The test is broadly applied in studies on gender (Merrit and Kok 1997; Daoud 1976) and in studies on violent behavior in prisoners (Lev-Wiesel and Hershkovitz 2000). This type of research using drawing has been attacked for lacking rigorous analysis (Webb *et al.* 1966) and it has been suggested that drawing methods are more valid and useful when aggregated with other data (Flanagan and Motta 2007). In anthropology, later research has gradually succeeded in debunking modern western claims to authority and superiority by focusing on comparative cross-cultural studies (Wilson and Wilson 1987; Alland 1980). It is better understood now that the creation of aesthetic representations in settings where people live in a traditional fashion can focus on universal patterns, local conventions or specific expressions in child and adult drawings (Anderson and Anderson 2009; Thompson 2008). They have merit standing alone in their own social context.

Practical Tips 8.1

TOURISM REALITIES

Tourism realities are paradoxical and favor:

- collaboration of researcher and respondent over authoritarian scrutiny;
- self-reflexive knowledge making over universal and absolute knowledge;
- interpretive and constructed narratives over empirical facts;
- context and perceptions over generalizations.

In tourism research the drawing methodology has been used to identify resident impressions of tourists (De Bruycker 2007; Gamradt 1995) and to identify resident and tourist perceptions of a destination (Hunter and Suh 2007), via its representation(s). These studies draw on a variety of theoretical perspectives and analytical methods. But they hold in common an approach that brings researcher and respondent into a partnership of meaning-making, recognizing the authenticity of the respondent's knowledge, unique identity, history and social perspective while requiring strict interpretive discipline on the part of the researcher. This approach is collaborative rather than authoritarian and findings are relativistic, local and nuanced. In this sense the drawing methodology along with other visual methods are timely and critically important to contemporary developments in tourism research that increasingly recognise that in tourism, realities are contingent and emergent (Stewart 2005).

The drawing methodology: an example

To better understand the drawing methodology in the context of tourism research, an example is discussed in this section. In a study already briefly mentioned (Hunter and Suh 2007), the effect of cultural commoditization on the perceptions of residents and visitors was explored. In this study an iconic cultural artefact of Jeju Island, Korea known as the "stone grandfather" or "dolhareubang" which is one of the most commonly recognized representations of traditional culture was the subject of research. There are only forty-eight of these stone relics scattered around the island (they are about 1.5–2.5 meters tall) but the stone grandfather image is routinely used in tourism promotion and advertising, or miniaturized and commoditized as souvenirs. The researchers wondered how tourists and island residents perceived the stone grandfather (as the original actually appears, as the stones are regularly represented in tourism media, or as they appear as souvenir commodities) and what effect common verbal references to the stones have on people who live on or visit the island. In other words, do commercial interests degrade the authenticity of cultural artefacts by distorting the public's perception of them?

To this effect, visual and verbal representations of these "dolhareubang" (or stone grandfathers) were collected and compared to originals. Then, a sample of people was recruited and asked to simply draw a stone grandfather, and to provide brief verbal comments describing the drawing. Drawings were processed and sorted to identify common characteristics and compared in turn to the visual and verbal variations of the originals and their representations. It was found that residents and visitors alike have set ways of perceiving and remembering this image. It seems that perceptions vary according to experience and that there is no single ideal or correct perception of this iconic cultural representation.

In Figure 8.1 a picture of the original stone grandfather is featured (left) with its various representations (right). Included (above right) are souvenir miniatures, souvenir key chains, and a stylized representation on a restaurant window. Also (below right) included are, public telephone booths, a banner for an international conference, and cartoon "mascots". As could be imagined, a visitor or resident is as likely to encounter a representation of the stones as they are to encounter the original.

Respondents were asked to draw the stone grandfather from memory. Each individual created an image that depicts certain features of the stones, and omits other features. In Figure 8.2 a selection of ten drawings of the stone grandfather reveals a number of very different perspectives. Of the ten drawings featured, five (above) were created by female respondents and five (below) were created by male respondents. The female respondents' drawings can be interpreted (from left to right on the top row) as representing: (1) an accurate depiction of the stone grandfather, including dots that represent the basalt stone from which it is carved; (2) a lack of knowledge of, or only a vague perception of the stone grandfather; (3) a depiction of the common verbal reference to the stones as a "kind grandfather"; (4) the common cartoon-like

Figure 8.1 The original stone grandfather of Jeju Island and its representations
Source: Author's own photograph, used with permission

Figure 8.2 Ten depictions of Jeju Island's "dolhareubang" as drawn by respondents
Source: Drawings created by respondents in the author's original research, used with permission

representation of stone grandfathers as "mascot"; and (5) a depiction of the common verbal reference to the stones as a phallic/fertility symbol.

The male respondents' drawings can be interpreted (from left to right on the bottom row) as representing: (1) the most common verbal references to the stone grandfather's most distinguishing features, its hat, "bulging eyes" and large nose (which women who wish to be pregnant are encouraged to touch); (2) a visually accurate depiction; and (3) a very limited knowledge of the stone grandfather. The two drawings on the far right are interesting because they are not accurate depictions of the "dolhareubang". Instead, they bear connotations of their authors' subjectivities. Their interpretation requires a deeper investigation into the respondent's subjective perceptions or attitudes. They could be considered as "statistical outliers" and may be nothing more than a sham on the part of the respondent . . . or evidence of a creative imagination that has supplied additional information to compensate for lack of perceptual experience concerning the image of the "stone grandfather".

It is clear that in all of these drawings, some of the participants are thinking of the "stone grandfather" objectively, as a "thing-in-itself"' and attempting to draw it exactly as it appears to them, from memory. Other participants are thinking of the "stone grandfather" metaphorically or as it has been verbally described to them and are attempting to represent these concepts as pictures. In other words, the choice to favour either visual perceptions or verbal accounts has a powerful effect on the ways in which a particular subject is rendered as a drawing. Each drawing reveals the "stone grandfather", isolated from its surrounding environment and depicting the preferred subjectivities of its author.

Taking these ten drawings as a whole, some additional observations can be made. It is interesting that in none of these drawings is there a horizon line or any other evidence that the respondents perceive the "dolhareubang" as a thing that exists in a physical context or environment. It is also interesting to observe that perceptions focus only on prominent features or key verbal elements such as the "stone", or the fertility motif that is commonly referred to in promotional literature, or the "kindliness" and "grandfatherly-ness" of these stone relics. For some respondents an accurate depiction of this cultural artifact might not have been possible due to lack of knowledge, lack of skill or a weak memory. Finally, a respondent's attitudes or imagination may contribute to a drawing that is so creative, or weird, that they overwhelm any interpretation.

Even in drawings that seem relatively similar, and drawn by similar respondents (in this case, females), there are differences. In Figure 8.3 note how each of the six drawings share some common characteristics, including the stone grandfather's hat, prominent eyes, nose and mouth, arms and legless torsos. However, there are significant differences, namely, the expression of the mouth, the placement of the arms and the rendition of the base. In drawings 1, 3 and 5 (numbered from left to right) the "dolhareubang" wears a smile. In drawings 2 and 4, the arms are correctly depicted, one hand above the other. In drawing 3, the important stone base upon which each stone is placed is not included in the drawing. It is up to the researcher in this or any case to determine what elements of respondents' drawings are to be considered and to what degree similarity or difference is to be defined. Due to the short time respondents have to draw, and the fact that they are, in this case, drawing from memory, these six drawings were considered to be of one category, that is, they were considered to be "similar enough".

This example of the drawing methodology, taken from research performed in Korea illustrates the complexity of the method and serves as a precaution that whatever respondents are invited to draw, the visual subject matter should be kept simple and clear parameters provided. Even so, respondents will tend to add their own creative embellishments, or they might fail entirely in the drawing endeavor. It is not recommended that respondents be asked to be artists. A cultural image as depicted by respondents is not a single thing, but

Figure 8.3 Similarities and differences between respondents' depictions of the "dolhareubang"

Source: Drawings created by respondents in the author's original research, used with permission

rather a fusion of literal and symbolic perceptions. Therefore, complex visual scenarios such as landscapes, cityscapes or detailed portraiture are not suitable objects of research.

But what a drawing shows about the respondent's relationship with a destination is highly dependent upon how the drawing is interpreted. To a significant extent interpretation of the drawing requires verbal cues and a social context for meaning. The age, gender, cultural background, socioeconomic conditions, education, nationality and other characteristics of respondents will have an effect on how they draw. The skill and level of detail in the drawing are related to education or training, attentiveness to the environment or the level of involvement with the research questions. The composition of the drawing (the arrangement of pictorial elements) reveals respondents' sense of priorities and importance. Thickness of line and certain embellishments reveal the certainty, anxiety or confidence of the respondent. In the drawing methodology, simple and clearly defined subject matter, clear theoretical goals and systematic analysis determine findings. The size of the research sample can affect the researcher's ability to generalize to the wider social context (a large respondent sample and limited elements in drawings) or to the specific sample of respondents (a small sample and more complex drawings). The drawer's level of skill or knowledge: how skilled is the respondent in drawing, diagramming or mapping; how well does the respondent understand, remember or perceive the contents and context of a destination and its features?

The drawing methodology: theory and practice

Having touched on the logic of drawing, its place in the social science research tradition and the example of perceptions and the Jeju Island stone grandfather, it would seem that a formal outline of the theory and practice of the drawing methodology should follow. In this section the key philosophical and practical principles of the approach and its advantages and pitfalls are

Practical Tips 8.2

DRAWING IS SELF-REFLEXIVE

A drawing is self-reflexive and can reveal:

- **What the drawer sees in an environment**: What significantly comprises/represents the destination's built/natural environment; what features are seen as cultural/urban/rural; what kinds of people inhabit the environment; how do they appear and what are they doing?
- **How the drawer judges the reality of an environment**: How are things arranged – clean/dirty; crowded/empty; scenic/ugly; what is liked/disliked/feared; what is important/unimportant?

Practical Tips 8.3

THE ANALYSIS OF DRAWINGS 1

In the analysis of drawings, the researcher must consider the respondent's self reflexivity:

- how well the drawing represents the subject matter (the place, person or thing that is identified as the subject of the research);
- how the drawing represents the individual who created it (the subjective and self-reflexive perceptions of the respondent).

described. Key issues in the total process are described, including the construction of research design, strategies in sampling, the development of research instruments, data collection, analysis and interpretation.

Research based on drawing has focused alternately on what the respondent can reveal about certain subject matter and, on what the drawing reveals about the respondent. In the first case the respondent is considered an informant equipped with certain unique perspectives that are valuable to the researcher. In the second case, the respondent is usually considered a subject who is distressed, impaired or pathological. In reality the success or failure of the drawing methodology depends on the ability of the researcher to employ good judgment and rigorous analysis and the recognition that perspective and pathology are inseparably entwined. The connotative meanings of the drawing are reflexive, that is, the representational meanings it conveys tell something about both the subject matter and its creator. These special characteristics contrast the drawing methodology and the standard verbal-response questionnaire driven research, or open and closed responses.

Research design

The development of a research plan using the drawing methodology must begin with a tourism problem that can be securely positioned in terms of theory, and practical policy, social or management concerns. Problems can be associated with cultural authenticity, commoditization, perception, satisfaction, sustainability or any number of popular research issues. A visual standard must be set of what the research object is (a person, place or thing), and how it appears in the real environment as an original, a representation or a commodity. Often, a characteristic of people, places or things are reproduced as representations (on postcards, in brochures or guidebooks, on signs or as souvenirs). There also may be various standard or common verbal descriptions and these are also representations of the research object. And the researcher must recognize that often visitors or residents encounter these representations more

Practical Tips 8.4

RESEARCH DESIGN

Research design should include:

- properly formulated theoretical themes that reflect social or technical issues;
- consideration of specific theoretical, policy or management concerns;
- a visual standard of what respondents might potentially depict in drawings;
- a standard as to what kind of participant or respondent is suitable to the research scenario.

frequently than the original object. In this sense, the researcher must be visually aware of how the object of interest might be perceived. Finally, a standard must be set as to what kind of participant or respondent is suitable to the research scenario.

The potential for a causal link between theory and the research object should be outlined or at least anticipated. The drawing methodology is used to elicit and demonstrate that link, and in this sense, the research object works as a barrier, catalyst or problematic entity that connects to theory, and in addition, to a particular population of individuals. The drawings are the primary data of concern and they are created by respondents. In order to make best use of the important role of the respondent, the researcher should be prepared to make research intentions and goals clear to respondents prior to drawing. Should a drawing be creative, or should it hold to certain conventions? Is the emphasis on the research object, or on the respondents' subjective perceptions? Will there be problems in using the media, or in the composition of the drawing?

The importance of sampling

Good sampling is critical to success in the drawing methodology. Sampling is the systematic process of finding people to act as respondents, to participate in the researcher's inquiry. In the drawing methodology, the research sample is a group of people specifically recruited to draw, and their numbers need not be large. Instead, they should be selected on the basis of their willingness to draw and their ability to provide intelligence concerning their drawings in verbal responses regarding the research problem. The interpretation of drawings is not intended to represent the perceptions of all people in a population, but to represent all the perceptions of the sample. In this sense, sample size does not necessarily have to be large. The drawing methodology

Practical Tips 8.5

PURPOSIVE SAMPLING

Use purposive (theoretical) sampling to:

- choose respondents who are able to understand your research purpose and goals;
- choose respondents who are involved;
- choose respondents who will draw in a way that can be effectively interpreted.

is not, by definition, either a qualitative method or a quantitative method. That distinction depends on the choices made by the researcher concerning the analysis, interpretation and presentation of findings. However, for most researchers a sample of less than, say, thirty participants would probably not yield enough depth and diversity of drawings to work with.

The researcher must make informed decisions in the generation of a *purposive* selection, or sample of people that will be representative of the researcher's inquiry rather than of the population. Purposive sampling (Noth 1990) seeks "a diverse non-random selection" (Fairweather and Swaffield 2001: 288) of respondents and seeks a maximum *variety* of perspectives (Patton 1990). Purposive sampling is favorable to the interactive knowledge making process (Gall *et al.* 1996) when it targets a sample that has concerns specifically relevant to the research problem (Brown 1980). Purposive sampling is favored in qualitative research (Lincoln and Guba 1985; Guba and Lincoln 1994; Glaser and Strauss 1967) where knowledge-making is interactive and dependent upon the expertise or authenticity of respondents' experience as much as it is dependent upon the researcher's interpretive skills. For the drawing methodology, respondents should be selected based on their proximity to the research problem, their ability to produce drawings that will be of value and their ability to respond verbally to researchers' questions.

On-site data collection and measurement instrument development

On-site data collection begins once a clear research design and sampling is complete. A researcher may decide to instruct sampled respondents to draw together, at one time and place or one-by-one as the sample grows (the snowball effect where one respondent introduces or recruits others). In either case the research tool or measurement instrument should be complete and should include: (1) a brief introduction to the research and instructions for

Practical Tips 8.6

SAMPLE INSTRUMENT FOR THE DRAWING METHODOLOGY

Preliminary introduction and instructions:

Name of affiliation and logo______________________________

Introduction, including title of research, the purpose, and a request for respondent's participation:

> "In this survey we would like to know more about your impressions of __________ . We hope you can take a few moments to reply to the following questions."

1 Checklist for respondent's basic descriptive information

(A) *Please answer these general questions first:*

General demographics	*Questions for the visitor*	*Questions for the resident*
Gender ________	Is this your first visit to ______ ?	How long have you lived in _______?
Age ________	How often do you visit ______ ?	Why did you relocate to _______?
Nationality ________	What is your purpose of visit?	How frequently do you interact with tourists _______?
Place of residence ________	• Tour ________	How would you describe your interactions with tourists: (e.g. business, similar activities, on the street, etc.)
Education ________	• Independent traveler ________	______________________
Occupation or affiliation ________	• Visiting friends or relatives ________	______________________
Annual income ________	• Business ________	
	What is your length of visit ______ ?	

(B) *Additional verbal response questions for respondents*

Please choose the best answer for each question, *or* please fill in the blank ______ .

Note

If this is a pilot study, there should only be a space for a few descriptive comments regarding the respondent's drawing, and it should *follow* after the drawing.

If this is a study based on pilot testing, a more developed hypothesis-driven questionnaire might be included and it should *precede* the drawing.

Practical Tips 8.6 *(continued)*

2 The respondent's drawing

Please draw a picture of ______________________________ as you remember/see/imagine it.

Please do not worry about your drawing skills – just use your own style to complete the drawing. Choose 'portrait' arrangement or 'landscape' depending on which best suits the subject matter.

Also, please add a few comments describing your drawing.

Note: It is best to provide a full page for the drawing.

the respondent; (2) a checklist for respondents' basic descriptive information; (3) a space (optional) for additional verbal response questions; and (4) a space for the respondent's drawing (clearly delineated with a border or frame).

It is appropriate for the researcher to be present, to ensure that the respondent(s) is/are clear on the expectations of the researcher, and so that the researcher can take note of the respondent's actions and behaviour. Clear verbal instructions, in addition to those written on the response form, are important and it should be specified whether a drawing, sketch, diagram or map is preferred. Ensure that respondents have a relatively uniform amount of time and minimal distractions and check to ensure that they do not cross-check each other before or during the drawing. Reassure respondents that their drawing and their points of view are unique and valuable. Empower them and let them have fun. It is not the artistic value of the drawing that counts, but its authenticity. Make explicit how the respondent is to draw: from memory, observation or as creative expression depending on the circumstances.

A pilot test may be necessary to ensure that respondents are able to draw what is expected of them. In this sense the drawing methodology is inductive, beginning with the preliminary triangulation of research goals, theoretical themes and an appropriate sample of respondents. An impressionistic record (fieldnotes) generated by the researcher on-site during data collection can contribute to better primary analysis. Glaser and Strauss (1967) would call this the "constant comparative method", or analytic induction.

Analysis

In this chapter, five general steps in the analysis of drawings are recommended: (1) collect and code; (2) generate descriptive statistics; (3) perform content analysis to generate drawing "types"; (4) generate statistical analysis to identify the validity and reliability of "types"; and (5) perform interpretive analysis using a general framework of semiotics and relevant theoretical themes. As the researcher develops a style or approach that is unique to the needs and demands of the research context these general steps might change. For example, sample size will directly affect the relative relevance or importance of "statistical" versus "interpretive" analyses. A quantitative analysis that stresses the generalizability of findings within the sample to a population would emphasize a larger sample whereas a qualitative analysis would be more concerned with what findings reveal about the respondents in the sample. Exploratory studies might also proceed with smaller samples.

Regardless of the approach, it holds that good collecting and coding is important to ensure the anonymity of the respondent (by substituting respondent name for a coded number) and accuracy of the findings (by developing a systematic and orderly codebook for analysis). Drawings once numbered can be sorted manually as-is or as transformed into digital media using camera or scanner.

Content analysis begins with the researcher's identification of drawings' hypothetical dimensions. Similarities and differences are sought that will differentiate "types" as categories that are mutually exhaustive and exclusive, or that hold together through internal convergence and external divergence (Guba 1979). When coded along with respondents' descriptive data, statistical factor analysis will add empirical support for "types" and simple one-way ANOVA will help with establishing relationships between drawing "types" and respondent "types". If statistically significant differences between drawings' dimensions are not found, the process may need to be repeated. The statistical analysis of drawing "types" is only one step in the research process and is meaningless without theoretically-based interpretation. In other words, a large sample does not guarantee successful results, and statistical consistency does not guarantee the reliability of a study. However, a large sample such as that used in the "stone grandfathers" study (Hunter and Suh 2007) of 179 participants gave the researchers enough data to make confident and informed interpretations. It is highly unlikely that a sample less than, say, 30 will yield the necessary breadth and diversity of drawings that represent all potentially held subjectivities.

Identify elements of the drawing's main subject, the clearest set of dimensions for primary variables. Additional elements of the drawing (point of view/perspective, foreground/background arrangements and proportional relationships) might offer additional information concerning respondents' attitudes towards themselves or the subject matter depending on the nature of the research problem. This process often requires several runs to get the best results. Careful analysis is important to avoid overlooking "hidden variables" or elements in the drawing that are unexpected but potentially meaningful.

Practical Tips 8.7

THE ANALYSIS OF DRAWINGS 2

The analysis of drawings includes five steps:

1 Collect and code.
2 Generate descriptive statistics.
3 Perform content analysis to generate drawing "types".
4 Generate statistical analysis to identify the validity and reliability of "types".
5 Perform interpretive analysis.

Interpretation

When statistically significant categories have been defined through analysis the researcher is ready for the work of interpretation. In the drawing methodology, it is up to the researcher to identify to what extent the drawings produced by respondents convey information regarding the object of research or information about the respondents themselves. Drawings are self-reflexive, revealing things about the respondents that can be both purposeful and subliminal (Malchiodi 1998). Good interpretations focus on seeking patterns with relevant dimensions, and an overall holistic and complex description. Dimensions or "types" might be statistically significant but irrelevant to theory or the practical research questions. In any case good interpretation is informed by theory that represents the drawing as a depiction of the research problem. Interpretation can be informed by:

- *Social conventions*: The respondent can make stylistic and compositional choices based on learned meanings and social or cultural conventions (physical relationships between objects, perspective and point of view, the influence of iconic images, cartoons, modes of popular art) in the social environment.
- *Power relations*: Power can be represented by literal depictions (drawings of policemen, teachers, politicians or other authority figures, industrial buildings, churches, male/female relationships, adult/child, tourist/resident) or by the composition (scale, proportion and proximity).
- *Metaphors*: Modes of representation in drawing can express something more than a combination of images only. They can express timeless

Practical Tips 8.8

DRAWINGS ARE PRE-LINGUISTIC SIGNS

Drawings as non-linguistic signs convey information in a different way to language:

- In fixed sign systems (languages), each sign has a relatively objective and arbitrary meaning or function; there is a formal grammar; and each sign is associated with an utterance. In this case, signs are denotative and absolute.
- In flexible sign systems (drawings), each sign may be imbued with the subjective cognitive and affective meanings of its creator in addition to any of a number of objective (representational) meanings; there is an informal grammar (horizon line, perspective, resemblance, contingency); and each sign may be associated with a name but not an utterance. In this case signs are connotative and require interpretation.

themes of the human condition (confrontation and isolation, dualism and separation, achievement and vain pursuits, exploitation and justice, pain and consolation, the spiritual and the profane).

- *Metasystems*: These seek connections between drawings and elements of environments they represent: landscapes, cityscapes, townscapes, cultural/ heritage-scapes, tourist attractions, parks, or beaches.
- *Style*: Consider hidden variables in the form of: thick or thin lines, solid or broken lines, shading, extra scribbling or doodling, placement within the frame, relation to other drawn objects, symmetry, color, tone, point of view, proportion and perspective.

Critical issues

The drawing methodology in tourism research is a practice in dualism, straddling the gap between (or joining together) objective reality and subjective perceptions. Respondents will include deliberate and subliminal elements in their drawings and so careful recruiting in the form of purposive sampling is important. Respondents will produce drawings that reveal their unique knowledge, interests and subjectivities. Drawing is a self-reflexive methodology that reveals things about how a particular type of respondent perceives or remembers tourism—people, places and things—and at the same time, how that particular type of respondent perceives self. Even in the analysis, the researcher must play a role in that reflexivity, using qualitative and quantitative tools to translate the visual sense of a drawing into the verbal language of an interpretive narrative. Statistical and content analysis are used to identify categories that describe the typology ("types") of drawings made by respondents, and interpretive analysis explores the relevance of drawings in light of the problems and theory that drive the research.

In tourism research, the drawing methodology depends on good research design rather than good drawings. Interpretation must regard the many facets of respondents' subjectivities and perceptions as no less "real" than the objects they have been asked to draw. Drawing is a form of visual research that seeks out the moments before the unavoidable linguistic construction of reality where signs and signifiers combine in cognitive symbolic relationships (Hirschman and Holbrook 1992: 29). Culture is a "context dependent semiotic system" (Jenks 1993: 121) and individuals "in" and "out" of culture hold a range of literal and symbolic perceptions that reflect the emergent qualities of social reality. Drawing gets to the root of visual perception as it is in the mind (and hand) of the respondent. However, good results in this style of research depend on attention to certain pitfalls in the drawing methodology, namely: (1) the problem of representation; (2) the problem of replication; and (3) the problem of responsibility. This chapter closes on a brief discussion of each and what can be done to increase the "trustworthiness" of this method.

The problem of representation

The drawing methodology seeks definable categories in a sample of depictions of a visual problem in tourism research. Content analysis is not a sufficiently rigorous approach to defining drawing "types" and requires statistical analysis to refine hypothetical categories. On the other hand, in statistical analysis the accuracy of the internal mathematics can be confused with the accuracy of the premises upon which they are based. There must be a balance of qualitative and quantitative specifications upon which the drawing is evaluated.

There is always a double threat of researcher and respondent bias. The researcher's bias arises when there is not enough self-reflexive awareness of the role the researcher plays in the collaborative generation of visual data. There may be gaps in understanding between the researcher's subjectivities and identity and those of the respondents' which will lead to the emergence of "hidden variables" or distortions in interpretation. The best insurance against hidden variables is to keep the drawing simple. Complex drawings decrease interpretive validity. Liberal projective diagnoses should not be offered in the researcher's interpretations. Respondent bias arises when the respondent deliberately shams the research or is not properly briefed before or during the drawing. The researcher should be vigilant for conflicts of interest and ready to adjust research goals, problems or theoretical claims accordingly. Respondent bias is exacerbated when their drawing skills are not developed enough to properly represent their perceptions or memories. Ensure, in on-site or post-research debriefings, that respondents have been comfortable with the task given them and that a true researcher-respondent collaboration has taken place. When possible, follow-up interviews or photo-elicitation methodologies can ensure greater representation in research based on the drawing methodology.

The problem of replication

The drawing methodology seeks trustworthiness with transparent research methods and interpretation. Only research that is replicable can be scientific and tourism researchers employing the drawing methodology should enhance the approach by checking for what researchers, including Gamradt (1995) call for, including:

- *Credibility*: Prolonged engagement, persistent observation and referential adequacy.
- *Transferability*: Congruence between purposive sampling and thick descriptions.
- *Confirmability*: Documentation of changes in the research plan using a reflexive journal and the use of an auditor or second opinion in data interpretation.

In the drawing methodology, factors or "types" might be paradoxical or entangled, but there are checks and balances to ensure that findings are credible and corroborative. Validity, referring to the problem of whether a study measures what it is intended to measure, is strengthened when statistical methods (basic factor analysis) are performed in tandem with content analysis, and further strengthened through pre-testing or pilot tests. A proven conceptual framework (Echtner and Ritchie 1991) can contribute to credible findings. Ideally, findings or interpretations should be compared to what emerges in follow-up interviews with respondents. In this sense, validity is intimately related to the problem of representation previously discussed.

Reliability of a study (its consistency or repeatability) is strengthened by corroboration with respondents (they are collaborators in the drawing method) to check for shamming and error or for instability between found "types." If findings do not make sense to the respondents, there is a problem in analysis or interpretation. Statistical consistency, as mentioned above, does not guarantee the reliability of a study. Drawing is a good method for exploratory work, and future research can replicate findings by recruiting new samples of respondents in the same context or in a different context and using the same conceptual framework. Cross-checking results should reveal whether or not the research question, conceptual framework and sampling have a good fit.

The problem of responsibility

The drawing methodology is sensitive to socio-economic power relations in tourism. A respondent may feel invested in a community or a cultural perspective and may also be concerned about certain destination development and profitability issues related to the local tourism industry. These two interests might be closely related or highly at odds. The sometimes contradictory interests of tourism make potential respondents highly sensitive to the practical value or contribution of tourism research, or the role of the researcher. Increasingly, residents of some communities are less and less likely to respond to surveys and questionnaires or interviews that seem to offer no real life contribution. Researchers in tourism must become socially conscious and active in their research by ensuring participant confidentiality while at the same time developing creative ways to debrief respondents, to make the research procedure educational and relevant to the community in which it takes place.

This becomes even clearer when the drawing methodology is employed as drawing can bring out emotions, attitudes, beliefs or other subjectivities. A responsible researcher must realize that findings are based on sensitive power-reflexive conditions and might arouse conflict. A good researcher should take time to return to the site to debrief respondents and let them know the results or interpretations or to solicit their opinions and expertise. The feasibility of an on-site post-research debriefing depends on sample size, but websites, blogs or email mailing lists are new and effective alternatives for disseminating findings and their implications and for taking tourism research out of the laboratory and exclusive realm of academics and into the community.

CHAPTER SUMMARY

The drawing methodology has been deployed in anthropology, the arts and the social sciences for decades with varying degrees of success. The possibilities in tourism research are still, however, largely uncharted. With strong theoretical development and attention to critical social and cultural issues mentioned in the chapter this visual approach can enable researchers to better appreciate how people perceive persons, places and things and perhaps more importantly, how they express their own (often hidden) subjectivities. In summary, five key points conclude this chapter:

- **Tourism realities**: The drawing methodology works best when there is a continuous and egalitarian collaboration between researcher and respondent. Dynamic and changing power relationships are the stuff of tourism and the researcher must be continually vigilant in order to empower respondents to recognize them and to prevent them from insinuating themselves into the research dynamics.
- **Perceptions and self-reflexivity**: While respondents attempt to depict what they see or remember, they are also consciously or unconsciously expressing their held subjectivities. The researcher must pay attention to how people perceive "realities" as literal, metaphorical or expressive. A drawing is a complex mix of skill and imagination where the environment and the self come together.
- **Drawings are pre-linguistic signs**: Drawings are enigmatic and deceptive when they appear without verbal anchors or context. The researcher must be sure to set an explicit visual standard (representations of a person, place or thing) to ensure that respondents know what it is that they are expected to draw. Photo-elicitation or onsite briefings and debriefings are crucial for good results.
- **The analysis of drawings**: Successful research using the drawing methodology depends upon good research design and theory, well-executed sampling procedures and good drawings. In the analysis of drawings there must be an informed balancing of statistical and interpretive procedures to ensure that all meanings, overt and hidden, are found.
- **Reliability and replication**: In the drawing methodology there will always be a double threat of researcher and respondent bias. There may be gaps in understanding between the researcher's goals and respondents' held subjectivities. Rigorous sampling procedures, statistical and interpretive analyses, and most of all a clear and well-defined subject for drawing are the only defenses against distorted interpretations and overlooked "hidden variables."

Annotated further reading

Crouch, D. and Lübbren, N. (eds.) (2006) *Visual culture and tourism*. New York: Berg.
This is a key text in tourism research that brings together two distinct cultural formations, "visual culture"' and "tourism" in a number of essays, or case studies.

Echtner, C. and Ritchie, J. (1991) "The meaning and measurement of destination image", *Journal of Tourism Studies*, 2: 2–12.
This is a classic journal article that highlights the notion of "destination image" in tourism research, a key theoretical theme in visual methods research.

Hall, S. (ed.) (2002) *Representation: cultural representations and signifying practices*. London: Sage.
This text deals, in depth, with the question of "representation", one of the central practices in the production and circuit of culture. It is required reading for anyone interested in researching tourism with visual methods.

Hunter, W. C. and Suh, Y. K. (2007) "Multimethod research on destination image perception: Jeju standing stones", *Tourism Management*, 28: 130–139.
This is the original journal article that describes the visual methods used in the research performed concerning the Jeju Island Standing Stones, the example discussed in depth in this chapter.

Metro-Roland, M. (2009) "Interpreting meaning: an application of Peircean semiotics to tourism", *Tourism Geographies*, 11: 270–279.
This journal article offers a simple and parsimonious introduction to Peircean semiotics and its relationship to tourism research. Semiotics has important theoretical and methodological relevance to research using visual methods.

References

Alland, A. (1980) *Playing with Form: Children Draw in Six Cultures*. New York: Columbia University Press.

Anderson, I. and Anderson, S. B. (2009) "Aesthetic representations among Himba people in Namibia", *International Art in Early Childhood Research Journal*, 1: 1–14.

Ashton, D. (1972) *Picasso on Art: A Selection of Views*. New York: The Viking Press.

Banks, M. (2001) *Visual Methods in Social Research*. London: Sage.

Berger, P. L. and Luckman, T. (1966) *The Social Construction of Reality*, New York: Anchor Books.

Brown, R. H. (1995) "Realism and power in aesthetic representation", in R. H. Brown (ed.) *Postmodern Representation: Truth, Power, and Mimesis in the Human Sciences and Public Culture*. Chicago: University of Illinois Press.

Brown, S. R. (1980) *Political Subjectivity: Applications of Q Methodology in Political Science*. London: Yale University Press.

Checkland, P. B. (1985) "From optimizing to learning: a development of systems thinking for the 1990s", *Journal of the Operational Research Society*, 36: 757–767.

Collier, J. and Collier, M. (1999) *Visual Anthropology: Photography as a Research Method*. Albuquerque: University of New Mexico Press.

Crouch, D. and Lübbren, N. (eds.) (2006) *Visual Culture and Tourism*. New York: Berg.

Curtis, G. (2006) *The Cave Painters: Probing the Mysteries of the World's First Artists*. New York: Alfred A. Knopf.

Daoud, F. S. (1976) "First-drawn pictures: a cross-cultural investigation", *Journal of Personality Assessment*, 40: 376–377.

De Bruycker, D. (2007) "Perception of local children towards tourists visiting Bruges", *Omertaa*, 3, retrieved on 18 August 2009 from www.omertaa.org/index.php?option=com_content&task=view&id=48&Itemid=63.

Dikovitskaya, M. (2006) *Visual Culture: The Study of the Visual after the Cultural Turn*. London: The MIT Press.

Echtner, C. and Ritchie, J. (1991) "The meaning and measurement of destination image", *Journal of Tourism Studies*, 2: 2–12.

Fairweather, J. R. and Swaffield, S. R. (2001) "Visitor experiences of Kaikoura, New Zealand: an interpretive study using photographs of landscapes and Q method", *Tourism Management*, 22: 219–228.

Finn, M., Elliott-White, M. and Walton, M. (2000) *Tourism and Leisure Research Methods: Data Collection, Analysis and Interpretation*. NY: Pearson Education Limited.

Flanagan, R. and Motta, R. W. (2007) "Figure drawings: a popular method", *Psychology in the Schools*, 44: 257–270.

Gall, M. B., Borg, W. R. and Gall, J. P. (1996) *Educational Research: an Introduction*, 6th edn. New York: Longman.

Gallarza, M. G., Saura, I. G. and Garcia, H. C. (2002) "Destination image: towards a conceptual framework", *Annals of Tourism Research*, 29: 56–78.

Gamradt, J. (1995) "Jamaican children's representations of tourism", *Annals of Tourism Research*, 22: 735–762.

Glaser, B. G. and Strauss, A. L. (1967) *The Discovery of Grounded Theory: Strategies for Qualitative Research*. Chicago: Aldine.

Guba, E. G. (1979) "Naturalistic inquiry", *Improving Human Performance Quarterly*, 8: 268–276.

Guba, E. G. and Lincoln, Y. S. (eds.) (1994) *Handbook of Qualitative Research*, London: Sage.

Guthrie, R. D. (2005) *The Nature of Paleolithic Art*. Chicago: University of Chicago Press.

Hall, S. (ed.) (2002) *Representation: Cultural Representations and Signifying Practices*. London: Sage.

Hirschman, E. C., and Holbrook, M. B. (1992) *Postmodern Consumer Research: The Study of Consumption as Text*, London: Sage.

Hunter, W. C. and Suh, Y. K. (2007) "Multimethod research on destination image perception: Jeju standing stones", *Tourism Management*, 28: 130–139.

Jenks, C. (1993) *Culture*, New York: Routledge.

Jung, C. G. (1968) *Analytical Psychology: Its Theory and Practice*. New York: Vintage Books.

Kandinsky, W. (1977) *Concerning the Spiritual in Art*. M. T. H. Sadler (trans.), New York: Dover Publications, Inc.

Lawrence-Lightfoot, S. and Davis, J. H. (1997) *The Art and Science of Portraiture*. San Francisco, CA: Jossey-Bass.

Leeds, J. A. (1989) "The history of attitudes toward children's art", *Studies in Art Education*, 30: 93–103.

Lev-Wiesel, R. and Hershkovitz, D. (2000) "Detecting violent aggressive behavior among male prisoners through the Machover Draw-a-person test", *The Arts in Psychotherapy*, 27: 171–177.

Lincoln Y. S. and Guba E. G. (1985) *Naturalistic Inquiry*. London: Sage.

Machover, K. (1949) *Personality Projection in the Drawing of the Human Figure*. Springfield, IL: Charles C. Thomas.

Malchiodi, C. A. (1998) *Understanding Children's Drawings*. London: Jessica Kingsley Publishers.

Martlew, M. and Connolly, K. J. (1996) "Human figure drawings by schooled and unschooled children in Papua New Guinea", *Child Development*, 67: 2743–2762.

Mead, M. (1953) *Growing up in New Guinea*. New York: Mentor.

Merritt, R. D. and Kok, C. J. (1997) "Implications of the people = male theory for the interpretation of the draw-a-person test', *Journal of Personality Assessment*, 68: 211–214.

Metro-Roland, M. (2009) "Interpreting meaning: an application of Peircean semiotics to tourism", *Tourism Geographies*, 11: 270–279.

Noth, W. (1990) *Handbook of Semiotics*. Indianapolis: Indiana University Press.

Paget, G. W. (1932) "Some drawings of men and women made by children of certain non-European races", *Journal of the Anthropological Institute*, 62: 127–144.

Patton, M. Q. (1990) *Qualitative Evaluation and Research Methods*, 2nd edn. Newbury Park, CA: Sage.

Peirce, C. S. (1931–58) *Collected Writings* (8 vols.), C. Hartshorne, P. Weiss and A. W. Burks (eds.), Cambridge, MA: Harvard University Press.

Perez, B. (1988) *L'art et la poesie chez l'enfant*, Paris: Ancienne Librairie Germer Bailliere.

Pink, S. (2007) *Doing Visual Ethnography*, 2nd edn. London: Sage.

Pritchard, A. and Morgan, N. J. (2001) "Culture, identity and tourism representation: marketing Cymru or Wales?" *Tourism Management*, 22: 167–179.

Rose, G. (2007) *Visual Methodologies: An Introduction to the Interpretation of Visual Materials*, 2nd edn. London: Sage.

Russell, D. B. and Ison, R. L. (2000) "The research-development relationship in rural communities: an opportunity for contextual science", in R. L. Ison and D. B. Russell (eds.) *Agricultural Extension and Rural Development: Breaking out of Traditions*, Cambridge: Cambridge University Press.

Stewart, K. (2005) "Cultural poesis: the generativity of emergent things", in N. K. Denzin and Y. S. Lincoln (eds.) *The Sage Handbook of Qualitative Research*. London: Sage.

Sturken, M. and Cartwright, L. (2004) *Practices of Looking: An Introduction to Visual Culture*. New York: Oxford University Press.

Thompson, C. M. (2008) "Action, autobiography and aesthetics in young children's self-initiated drawings", *Journal of Art and Design Education*, 18: 155–161.

Thurman, J. (August 21, 2009) "First impressions: what does the world's oldest art say about us?" *The New Yorker*, retrieved on 21 August 2009 from www.newyorker.com/reporting/2008/06/23/080623fa_fact_thurman.

van Leeuwen, T. and Jewitt, C. (eds.) (2001) *Handbook of Visual Analysis*. London: Sage.

Webb, E., Campbell, D., Schwartz, R. and Sechrest, L. (1966) *Unobtrusive Measures: Nonreactive Research in the Social Sciences*. Chicago: Rand-McNally.

Wilson, B. and Wilson, M. (1987) "Pictorial composition and narrative structure: themes and the creation of meaning in the drawings of Egyptian and Japanese children", *Visual Arts Research*, 13: 10–21.

Wolcott, H. F. (1995) "Making a study 'more ethnographic'", in J. F. Maanen (ed.) *Representation in Ethnography*. London: Sage.

Part 4

Analysis and representation

9 Readings of tourist photographs

Michael Haldrup and Jonas Larsen

Introduction

Visual methods and materials, such as drawings, diaries, photographs and video footage, has proliferated in tourist studies recently. Visual representations produced by researchers and tourists provide rich and vivid sources for reading the 'lives' and practices of tourists. The proliferation of visual methods and materials raise the question of 'reflexivity'. As visual anthropologist Pink (2003) argues, there is no 'methodological fix' to avoid reflexivity. 'Reflexivity' revolves around the idea that field accounts and readings are just as much an account of researchers' gender, age, ethnicity and theoretical lenses as of the field and text/image. Reflexivity implies acknowledging that researcher and field or image/text are inevitably 'entangled' (Ateljevic *et al.* 2005) and mutually constitute each other in the production of 'knowledge'. More broadly, 'reflexivity' emphasises situatedness, embodiment and context (see Feighey 2006 for a review of reflexivity within tourist studies).

This chapter discusses different methodological and theoretical approaches for reading tourist photographs. Despite the fact that photography is an emblematic tourist practice and tourist studies have been dominated by a visual paradigm of gazing and photographing (Urry 1990), few studies have undertaken readings of photographs produced by tourists or made ethnographic visual material about practices of doing tourist photography. Retrospectively and reflexively, we discuss different readings by drawing upon our published studies of tourist photographs produced by ordinary tourists as well as our own visual ethnographic materials of picturing tourists. In these studies (see in particular Haldrup and Larsen 2003, 2010; Larsen 2005), we have undertaken different readings of tourist photographs, and in this chapter we hope to demonstrate the 'productiveness' of combining different theories and methods when 'reading' tourist photographs. Following Pink (2003), we argue for the usefulness of interdisciplinary approaches and we elucidate how reflexivity travels tacitly through contingent and often chaotic research processes in which different (contradictory) methodological tools may be used for making sense of tourist photography and visual material more broadly.

The first part of the chapter examines *representational* readings of the photographs, and we do so by discussing how we combined quantitative and qualitative readings of the subject matter of photographs. Drawing on recent discussions on materiality and the social life of photographs, we discuss *non-representational readings* of photographs as mobile objects and touchstones of memory. Here we turn our gaze from the content, codes and meanings of photographs, to their changing materialities and mobilities in time and through space. Through this twofold reading the chapter shows how the different conceptual lenses produce very different readings, which are not reducible to technical issues but also, as we discovered over time, induce significant differences in how we conceive of tourism, tourists and the places of tourism.

Representational readings

When we – together with Jørgen Ole Bærenholdt and John Urry – in 2000 began working on the project *Destination Construction and Development in Demark*, tourist studies were dominated by a 'visual paradigm' that prioritised representational approaches and methodologies; and our research was initially framed by John Urry's theory of *the tourist gaze* (Urry 1990). In this account tourism essentially is a 'way of seeing' and its pleasures thus grounded in the eye, in the enjoyment of 'consuming places' (Urry 1995) visually and semiotically.

One of our research questions revolved around examining how tourists consume and represent tourism places, and more broadly, how consumer preferences and practices contribute to the construction and development of tourist destinations. Our research took place in two Danish tourist destinations – the Island of Bornholm and the coast of Northern Jutland – renowned for their aesthetic scenery and the many urban Danish families vacationing there. We undertook a series of qualitative methods, such as participant observation, interviews and readings of tourists' own photographs collected during our fieldwork. We had a 'burning desire' to investigate how tourists make pictures, what they take pictures of, and how we can read such personal photographs meaningfully. And, equally significant, as cultural geographers and sociologists with an interest in space and place, we were also eager to explore what we could learn about tourists' consumption of places by reading their representations of them. To answer such questions, we collected tourists' photographs during field trips and 'read' these together. Furthermore Jonas Larsen undertook ethnographic observations and interviews with picturing tourists at one of the destinations.

At Bornholm we collected them by asking the interviewees for a copy of their films taken with their own cameras and ten families subsequently provided us with their photographs. We stressed that all photographs were of value to us, and that they should only thrust aside those that they were uncomfortable with being read and published. In Northern Jutland we offered people when arriving at the information office a disposable camera so they were aware from the start that their photographs would be read by us. And since we developed the films, people did not have a possibility to leave out images.

During our first round of field work, we slowly realised a discrepancy between our expectations, of observing and talking to urban families visiting and photographing renowned sites, known for their aesthetic qualities or 'difference', and the 'reality' of 'introvert' families having a nice time together, engaged in various mundane and somewhat ordinary activities, including photographing family members as much as attractions.

Somewhat puzzled, we decided to investigate systematically the *content* of the collected photographs to get an idea of what sort of motifs animate camerawork and the memories tourists bring back. Therefore, a quantitative content analysis became the first step in our reading of the many collected photographs. In the main text book of content analysis (1980), Krippendorf defines content analysis as an 'empirical grounded method, explorative in process and predictive . . . in intent' (2004: xvii). Content analysis relies on a positivist and naturalistic ontology that conceives of cultural texts as more or less mirrors of the world. Content analysis is a representational methodology aimed at providing 'convincing readings of cultural texts, and to draw various conclusions from them, by looking at the texts themselves rather than at the ways in which people actually consume these texts' (Slater 1998: 234). Content analysis is useful to identify 'patterns that are too subtle to be visual on casual inspection and protection against an unconscious search . . . for those [photos] that confirm ones initial sense of what the photos say or do' (Lutz and Collins 1993: 89). So we turned to this method because it allows a systematic method for 'picturing' the places, events, objects and people that these tourists pictured often and rarely.

Content analysers have developed a variety of techniques for securing consistency and accuracy in the categorisation (categories must be exhaustive and mutually exclusive) and coding (inter-coder reliability) (see Bell 2001 for such techniques). Content analysis is best suited to explorative studies using relatively simple counts of simple (preferably unequivocal and explicitly defined) variables to test heuristically formulated hypothesis. A content analysis has to be built around a few hypothesis and unequivocal criteria for categorising content.

Content analysis does *not* bypass reflexivity but dislocates it to the categorisation and coding process. The categorisation of the content of a photograph requires interpretation of what it actually shows. This can hardly be accurately defined without engaging properly with the interpretation of what we see. In our study we ultimately had to introduce a theoretically informed framework of 'performance' categories to arrive at a robust set of categories. In the words of Rose (2001), we count what we *think* we see.

Our thesis was informed theoretically by the 'tourist gaze' and the idea that tourists largely gaze upon and photograph places that they already consumed in image-form before the departure. Tourists are not so much framing and exploring as they are framed and fixed:

> Involved in much tourism is a hermeneutic circle. What is sought for in a holiday is a set of photographic images, which have already been seen in tour company brochures or on TV programmes. While the tourist is away, this then moves on to a tracking down and capturing of those images for oneself. And it ends up with travellers demonstrating that they really have been there by showing their version of the images that they had seen before they set off.
>
> (Urry 1990: 140)

This hermeneutic circle turns the photographic performances of *tourists* into a ritual of 'quotation', and it made us come up with the heuristic thesis that tourists in countryside regions tend to produce pictures of idyllic, rural landscapes devoid of human interference. We then established a framework of categories loosely inspired by the theatrical metaphors associated with the 'performance turn'. We made a series of content analysis – for example, of the 'scenes' (places) and 'actors' ('locals', 'other tourists' and 'friends/family members') (see Table 9.1).

A content analysis, with strictly defined variables, enabled us to produce a robust pattern in the sample. In line with our thesis, 'rural landscape' is most frequently pictured. It accounts for around one-fourth of the images. Given the vitality of attractions in much tourism theory, and both destinations have well-known cultural sites, the relatively low number of such pictures is striking. Noticeable, and contra our thesis, is that a wide range of stages are

Table 9.1 Photographic scenes (in %)[a]

Site/actors[b]	*Family members*	*Locals*	*Other tourists*	*No people*	*Total*
Rural landscapes	9	–	–	17	**26**
Residence	13	–	2	2	**17**
Beach	11	–	–	3	**14**
Cultural sights	11	–	–	3	**14**
Amusement parks, zoos, pool areas	7	–	1	2	**10**
Urban landscapes	7	–	–	1	**8**
Means of transport	3	–	2	1	**6**
Restaurant or other small facility	4	–	–	–	**4**
Museums and galleries	–	–	–	1	**1**
Total	**65**	–	**5**	**30**	**100**

Notes

a Values in percentages with values less than 1% not mentioned

b Only in foreground of the picture.

frequently employed and not all are typical objects of the 'tourist gaze'. Most striking is that 'residence' comes second. More photographs are taken at the second home, cottage or hotel than at attractions.

With regard to the 'actor' dimension of the pictures, we can see that more than half of them portray one or more family members or friends in the foreground, and this is in stark contrast to the low number of other 'tourists' and 'locals'. It seems that other people are avoided even when picturing at crowded attractions. Considerably less than half of them picture deserted 'stages'; being scenes worthy of portrayals on their own, which were lower than expected. While endlessly reproduced in postcards and brochures, the classical tourist image – a romantic picture of a deserted 'cultural sight' or 'rural landscape' – only accounts for a quarter of all the photos. A content analysis is a valuable tool to produce a crude overview of how frequently a particular element occurs, and Table 9.1 effectively demonstrates that much tourist photography revolves around capturing familial social relations. Yet our content analyses also had its limitations. A content analysis is only concerned with registering manifest surfaces. It cannot explain the frequencies of the patterns nor 'make sense' of them. It relies on discriminating between qualitatively different variables and not, for example, different degrees and the meaning of various blends/variations in the composition. For example, Table 9.1 shows us that these tourists do not seem to reproduce 'postcard motives' as much as we expected. A content analysis is necessarily mute about why this is the case. Nor can it say anything about why touring families are keen to photograph each other or what such photographs mean to them.

Therefore, to unveil some of the cultural and personal universe of photographs and to 'hear' tourists 'readings' of their own photography practices, we turned to qualitative interviews with 'picturing tourists'. What struck us in these interviews was the degree to which the interviewees accentuated desires for and energy in making family-oriented photographs. Many told us that they photograph in order to take their personal photographs, not just postcards, the latter being too impersonal and dead. Despite praising it for its beauty, some tourists denounce Hammershus as being unworthy for family snapshot backdrops; while shooting there they do not take photographs of the castle as such. As one said:

> No, we haven't taken photographs of Hammershus, we have taken atmosphere pictures where the family is at the centre, you can see that it is holiday, and Hammershus is in the background. But it [Hammershus] shall not fill the image, it is the family that shall fill it, right? And then the little memory of where we are. That has to be in the background.
>
> (Quoted in Haldrup and Larsen 2003: 31)

Neither the castle's symbolic aura nor its romantic grandeur makes the camera work in awe. That is, unadorned by familial faces. 'Loved ones' have to enter the 'picture' to attract and energise camera action. Respondents were interested in making photographs that explicitly connote a holiday feeling. Being a recognised tourist site, Hammershus fits that role, but so do numerous other 'stages', even those less visually spectacular such as the place of residence.

These interviews provided explanation to the finding in the content analysis that 'familial faces' take the centre stage in many of the photographs. And it became increasingly clear to us that family photography and tourist photography are intricately intertwined, and we now began 'reading' tourist photography through the lenses of the sociological literature on 'intimacy' (e.g. Giddens 1992; Beck and Beck-Gernsheim 1995; Chambers 2001) and 'family photography' (Spence and Holland 1991; Kuhn 1995; Hirsh 1997; Holland 2001) rather than the tourist gaze and other tourism theory. The former helped us to understand family life as an accomplishment of performances, and the latter that the 'home' of 'ordinary' photography is the mobile and 'private' modern family and that it works through personal geographies of emotions, desires and memory: through a discursive economy of what we came to term 'imaginative families'. Families are not acting for the camera in a socially non-discursive vacuum; they respond to dominant mythologies of family life, to conceptions they have inherited, to images they have seen on television, in advertising, in film. While a private performance with its own dynamic play of gazes and roles, a profound need for producing 'correct' images is displayed, crying out that our family is an affectionate group because the ideology of family life teaches that the natural state of familial being is unity and tenderness. So gradually we came to think of tourist photography as 'part of the theatre that the family constructs in order to convince itself that it is together and whole' (Krauss quoted in Smith 1998: 16; Hirsch 1997: 7). Instead of understanding tourist photography as an essentially visual phenomenon that unambiguously feeds the power of the 'tourist gaze' and transforms lived places into fixed and identical sights, we began grasping practices, objects and places of photography as embedded in time-spaces of self-(re-)presentation and narration of family life.

The work of cultural geographers of the mundane practices of tourism and landscapes (e.g. Crouch 2003) was another theoretical lens that started guiding our readings. Tourist studies have struggled to account for the thick sociality and 'banality' of most tourist practices; it has been drawn to the spectacular and exotic, thus excluding more mundane types of tourism such as family vacationing in summer cottages and resorts. Much family tourism is about being together as *one* social body, to be face-to-face in an era of fragmented and liquid family life (Bærenholdt *et al.* 2004; Haldrup and Larsen 2010).

These literature reviews and our empirical 'findings' made us construct the notion of the 'family gaze' – a notion informed by, yet also departing from, our formative theoretical lens of the tourist gaze – to capture how family photography is socially organised and systematised in family tourism (Haldrup and Larsen 2003). The family gaze brings questions of sociality and social relations into discussions of tourist vision and photography. While Urry's gazes are mainly directed at extraordinary material worlds, the family gaze is concerned with the 'extraordinary ordinariness' of intimate social worlds. Yet material places are not unimportant to this vision. Rather, it performs places differently from the other gazes: places become scenes for acting out and framing active and tender family life for the camera. Family members and their performances make experiences and places extraordinary and full of enjoyable life.

Subsequently, we read *single* images that we regarded as typical of the family gaze. Here we turned to the qualitative methods of semiotics (Barthes 1977; van Leeuwen 2001) and cultural studies (Lutz and Collins 1993; Lister and Wells 2001). In contrast to the step-by-step world of content analysis, these are imaginatively and explicitly theoretically informed: they do not offer guidelines but clues and ideas. They can help to identify the dominant cultural codes governing and constituting such tourist photography. Since photographs are polysemic it follows that there can be no absolute readings, only partial ones mediated by specific 'ways of seeing'. Rather than discovering their meanings the researcher's readings partly construct them.

Such semiotic and cultural readings may be helpful in exposing some of the significant cultural *codes* – the implicit collective rules – that govern photographic performances of the family gaze. Our readings suggest that 'the family gaze' is put to work to portray idealised family relations and public places that nonetheless appear natural and private. And we exposed some of the codified nature of tourist photography's naturalness: photographic events

Figure 9.1 Family gaze photograph

Source: Research participant's photograph, used with permission

are highly staged and performed in accordance with a small number of scripts – codes. The loving family is fashioned through eye contact (see Figure 9.1), entwined bodies, playing bodies and solemn bodies. Embracing such conventional codes enable families to contain the tension between the longed-for ideal and the ambivalence of lived experience (Holland 1991: 4). This way of reading the photograph significantly departed from the inherent positivism that sustained our quantitative readings. Rather than viewing photographs as mirrors of the world, we came to view the production of the photographic object as part of the machinery that families draw on to *construct* their world. Hence, this turn towards a postpostivist and constructivist ontology framed the way we came to read each of the photographs as 'emotional objects'. However it must be emphasised that retrospectively it is notable that the way we read people's holiday photographs did not escape reflexivity. Rather the quantitative and qualitative techniques dislocated the crucial role of the researcher from the situated interpretations at play *within* the field. With regards to our content analysis this reflexivity migrated to the coding procedures and techniques involved in establishing rigid and unambiguous categories after the collection of photographs was completed. Within the semiotic readings of emotions and feelings we found embedded in the photographic objects the researchers' personal presumptions clearly shaped these readings but were not directly mobilised in the analysis made. Instead, the issue of reflexivity travelled tacitly through our readings, without being explicitly addressed.

Non-representational readings

So far we have thought of photographs as signifying texts, and through representational readings we have tried to interpret their meanings and messages. Yet in our research on the family gaze and especially in later publications (Haldrup and Larsen 2010) we have also experimented with what we might term *non*-representional readings of tourist photography. Inspired by Thrift (2007), we may suggest that non-representational readings are concerned with photographs as material (digital) and animating *objects* and affective and corporeal 'readings' of photographs. Whereas the geography of photography used to be dominated by a representational paradigm, some geographers now explore the material and non-representational qualities of photographs (and images more broadly) as well as the embodied practices and performances of making photographs. In the late 1990s, Crang (1999) urged geographers to explore the eventful, corporeal practices that produce – give birth to – tourist photographs. We adopted such an ethnographic framework in a series of more-than representational ethnographies of picturing tourists in action – instructing, framing, shooting and posing – and theoretical discussions of tourist photographing as an embodied performance. We show that the 'hermeneutic circle' overlooks the embodied 'social dramas' of tourist photography. Our ethnographies show that when faced by the 'camera eye', people make other bodies for themselves. Activities and walking are put on hold, and in posing, people present themselves as desired future memories. Reflecting that photography generally does not so much reflect geographies as produce them, *new* bodies and 'ways of being together' were constantly produced when camera action began. In accordance with the late modern cultural code that tenderness and intimacy epitomise blissful family life, families act out tenderness and intimacy for the camera and one another (Bærenholdt *et al.* 2004: 105–24; Larsen 2005).

Another example is Rose's research on family photography that is concerned with doings and objects of photography; not so much with the birth of photographs but with their afterlife, as photographic objects in time and space. She partly wants 'to think of family photographs as objects because things are done to objects' and partly because 'while objects may be acted upon, they are far from inert' but in part constituted through their use and circulation (Rose 2003: 8). She examines qualitatively what middle-class mothers *do* – physically, emotionally and socially – with their photographs, to make them prompt memories, being exhibition-able and capable of reaching absent others.

Contrary to representational readings of content, codes, meanings and so on, a non-representational approach explores the roles that photographic objects play in the different social, spatial and cultural contexts they travel through and set more or less roots in. This means examining their more-than-representational, more-than-textual, more-than-human qualities and acknowledging a photograph's own agency. The notion of non-representation may mislead. As Lorimer argues, we should better speak of studies of the 'more-than-the-representational' (2005) as non-representational theory does not neglect representations but rather emphasises that representations are produced and act in themselves. And photographs are themselves 'blocks of space-time' that have spatiotemporality beyond the spatiotemporality of the referent it carries. As Latham and McCormack (2009: 2) say:

> an image is never just a representational snapshot; nor is it a material thing reducible to brute object-ness. Rather, images can be understood as resonant blocks of space-time: they have duration, even if they appear still. Furthermore, the force of images is not just representational. Images are also blocks of sensation with an affective intensity: they make sense not just because we take time to figure out what they signify, but also because their pre-signifying affective materiality is felt in bodies.

So non-representational geographers bring forth the fact that photographs are more than just representations, and while photographic *images* are caught up with the moment, photographic *objects* have temporal and spatial duration; they are performative objects and their affective sensations are bodily felt. Thus, more-than-representational theories of photographs are concerned with the performative agency of photographs as well as their emotional and affective qualities, something poignantly captured in *Camera Lucida*. In this piece, Barthes theorises photography with, and through, his own lived body and a few private photos, including one of his deceased mother. Photographs are affective objects that have the sudden power to animate the 'feeling body', with pleasure, love, grief, despair and hate: 'suddenly a specific photograph reaches me; it animates me, and I animate it' (2000: 20).

When reading photographs we often ignore the fact that photographs are not only two-dimensional images, but also three-dimensional things, often with a lasting afterlife. Photographs are something more than just representations: they are always both texts *and* objects. The object-ness of photographs is evident in Roland Barthes' account of the photograph of his deceased mother:

> The photograph was very old, the corners were blunted from having been pasted in an album, the sepia print had faded, and the picture just managed to show two children standing together at the end of a little wooden bridge in a glassed-in conservatory, what was called Winter Garden in those days.
> (Cited in Edwards and Hart 2004: 1)

Following such work, we began to think of how photographs are 'made, used, kept and stored' and how 'they can be transported, relocated, dispersed or damaged, torn and cropped and because viewing implies one or several physical interactions' (Edwards 2002: 67–8).

In our research on the family gaze 'materiality' occurred in relation to memory. In the interviews the theme of 'memory' cropped up persistently. One interviewee explained that photographing was 'very important: no pictures, no memory almost. Some memories but they fade very quickly' (quoted Haldrup and Larsen 2003: 38). Others emphasised that photography 'stopped' or 'froze' time: 'I think [photography] is an attempt to fix time, some experiences, a passage of time that has meant something to you, which you will like to hold on to as a memory' (ibid.: 39). 'Photographs', as material anthropologists Elisabeth Edwards says, 'belong to that class of objects formed specially to remember, rather than being objects around which remembrance occurs' (1999: 222). By materialising ephemeral events the photograph solidifies memories and translates them into other spatial and temporal contexts, such as the home. Like other memory objects, such as souvenirs, some photographs become material touchstones of memories; artefacts that trigger memories and stories calling for emotional responses and mind travelling. The interviews revealed that tourists' camera work rest on a desire to turn instantaneous and ephemeral tourism experiences into a life of eternity, to construct long lasting, tangible and personal memory stories. Thus, we came to conceive of tourist photography as a form of 'memory work' (Bærenholdt *et al.* 2004: 105–24); we took this as another explanation of the fact that our respondents portrayed their loved ones in homely surroundings or doing mundane, everyday like things (see Table 9.1 and Figure 9.1).

Being both the subject and object of the photographic event, the photographing families performed photography, and the images they produced, as a way of making sense of themselves through accumulating personal knowledge and memory stories. Instead of viewing photographs as 'mirroring' places and events we became increasingly aware of the social function of the photograph, as a social object carrying and triggering memories: 'touchstones of memory'. Photographs have 'the effect of bringing the past into the present and making past experience live' (Morgan and Pritchard 2005: 41). While photographs are appreciated for bringing back pleasant memories, they can always strike back and haunt us with unpleasant emotions (Rose 2003). Photographs are ambivalent and slippery objects of emotions, and they can trigger very different emotions and effects over time. While photographs are objects and images of the past, how we relate to and are affected by these traces of our former lives are always about today (Kuhn 1995: 16).

In our recent book *Tourism, Performance and the Everyday: Consuming the Orient* (Haldrup and Larsen 2010: 122–83) we developed these ideas into a series of *non*-representational readings of how digital photography transforms photographic objects as well as the production and consumption of photographs (see also Larsen 2008). In addition to ethnographic research of mainly Danish tourists in Turkey and Egypt, we visited tourists in their own home after a recent trip to Egypt and Turkey to explore the *afterlife* of their photographs. Often, the uses of photographs take place at home where photographs are stored, worked upon, displayed, distributed, shown, triggering travel tales and recalling memories. Hurdley argues for using home ethnographies to explore how 'material culture display practices in the home are ongoing accomplishments, bound up with negotiations of space time and identity' (2007: 355; see also Miller 2001) and especially how visual material such as photographs are used to personalise and communicate 'past' memories. So we observed how photographs are used, worked upon, filed, discarded and distributed and people showed us around in their homes, pointing out where and how they store, display, circulate and use their photographs. We looked at and talked about photographs contained in, worked and exhibited upon walls, frames, fridges, photo albums, shoeboxes, jewellery boxes, digital photo albums, mobile phones, cameras, computers, CD-roms, DVDs, email accounts, blogs and social networking sites (e.g. Facebook). So we moved around in their houses while observing and interviewing, and conversations revolved around photographs and screens at hand. Together with our respondents, we 'excavated' some of the afterlife of tourist photographs and more broadly the multiple ways in which the performances of tourism integrate with their everyday domestic spaces. Drawing on auto-ethnographic work we had done around the same sites as our informants had visited as well as our own experiences with taking, using and distributing photographs we were able to build up a reflexive account of how people and photographs were cohabiting the domestic spaces of kitchens, living rooms and studies.

The aim with this research was to contribute to the more-than representational readings of photographs, as affective, material and mobile objects, by exploring, conceptually and ethnographically, the complex lives of digital tourist photographs, in part by comparing and contrasting the life of analogue and digital photographs. Drawing upon the new mobilities paradigm within the social sciences (Urry 2007), we paid particular attention to mobilities of photographic objects. So this research 'tracks' photographs' spatialities and temporalities, their physical and digital materialities, (im)mobilities and 'placing' within and beyond 'networked households'. Here we argue that in order to explore the life of photographs we need to think, metaphorically speaking, of photographs as corporeal, travelling, ageing and affective 'humans' rather than conventionally as bodiless, timeless, fixed and passive images.

We acknowledge that photographs as objects sets them in motion – in space and time. Private photographs are often set in motion to bridge distance, emotionally reaching out for close ones at a distance. Particularly, we examine how digitalisation and 'Internet-isation' of photographs means that they travel much faster and cheaper by email than by mail. Emails, it seems, are 'born' to travel and compress spatial distances: they are indifferent to distance and number of destinations; they travel equally fast and cheaply to distant and many destinations as to near and single ones. We discuss and show ethnographically how our respondents easily (re)distributed tourist photographs to significant others at-a-distance and exhibited them on home pages, blogs and social networking sites.

Through such ethnographic readings we outline some general features of the life of digital tourist photographs. First, we show the significance of the screen; the camera screen is where most photographs are inspected immediately after springing into life as well as during their 'early days' (before 'uploading'). And if they survive deletion at this stage they might end up being uploaded to a computer, entered on databases and viewed on yet another screen: the computer screen. From here, a small selection are mobilised and distributed, emailed to email boxes or uploaded to social networking sites where they (hopefully) will be consumed on the right computer screens around the world. Whereas tourist photographs used to be fixed material objects with a secure stable home in the bookshelf-residing photo album, most are today variable digital objects facing unpredictable afterlives in computer trash bins, folders, email boxes, blogs and, increasingly, on social networking sites. This illustrates how we increasingly consume holiday photographs without necessarily being face-to-face with the photographer, and the lives of photographs become much distributed and public. And many photographic images now live, for shorter or longer periods, virtual, digital lives without any material substance – in cameras and computers as well as on the Internet. Moreover, some photographs have complex biographies, because they materialise, dematerialise and rematerialise, take and retake various forms and inhabit different materialities over time. And their corporeal and facial look is also potentially transform-able as the 'computer-hand' has the ability to reach into the guts of a photograph. Analogue photographs are inescapable both images and objects but this is not the case with digital photography. While camera screens have a material tactility, the photographs they display are images, not physical objects.

What is certain is that tourist photographs have become more visible, mobile and tied up with everyday socialising on various networked screens. And we may add, disposable. Lack of an 'aura of thing-ness' partly explains why so many digital photographs are short-lived – but also why they are valued as a fast mobile form of pictorial communication. Yet this also means that once a photographer lets loose a photograph on the Internet, (s)he loses control over its destiny, as friends or strangers may use it in unforeseen contexts or distribute it even further. As copy-able and timeless travelling bites of information, Internet-residing photographs face very unpredictable lives with multiple possible paths, and some of these may be potentially harmful and unpleasant. In this sense we have shown that the life of digital photographs are complex and unpredictable, and that their 'bodies' are mutable. Their life is not determined and it is a hard job to track their paths and predict their future whereabouts and characteristics, not to speak of age or expected life age.

CHAPTER SUMMARY

Drawing upon our own readings of tourist photographs, we have suggested a variety of methods that students and scholars can employ if one is to undertake a reading of tourist photographs.

- There is no methodological fix to escape reflexivity when doing studies of visual material. The gaze of the reader (researcher) is always framing and co-constructing what is read.
- It is not preferable to choose one single methodological approach. Rather an interdisciplinary approach employing various different methods throughout one's study may help open up more creative, more inspiring readings of visual material.
- Quantitative and qualitative approaches are not necessarily mutually exclusive. We believe that our 'representational' readings have gained richness by combining quantitative and qualitative textual readings as well as conducting interviews with the 'producing' photographers of them.
- Studies of visual material and culture may benefit from 'non-representational' readings where the focus is upon reading the agencies and affective qualities of photographs and how they travel across various screens and materialities.

Annotated further reading

Bell, P. (2001) 'Content analysis of visual images', in T. V. Leeuwen and C. Jewit, *Handbook of Visual Analysis*. London: Sage.

This chapter gives a good introduction to the techniques of coding and interpreting in relation to content analysis, and exemplifies the strengths and weaknesses of content analysis in relation to a study of images done by the author. Provides a good 'hands-on' manual to how to perform content analysis.

Haldrup, M. and Larsen, J. (2010) *Tourism, Performance and the Everyday: Consuming the Orient*. London: Routledge.
This book demonstrates how a 'non-representational' approach to the production, circulation, distribution and use of tourist images opens up a broader understanding of how visual practices in tourism relate to the everyday material cultures. In particular the book demonstrates how a variety of ethnographic methods may inspire studies of visual material and everyday life.

Pink, S. (2003) 'Interdisciplinary agendas in visual research: re-situating visual anthropology', *Visual Studies*, 18 (2): 179–92.
This article considers advantages and pitfalls of interdisciplinary approaches in visual studies, with a particular emphasis on the role of reflexivity in visual anthropology.

Rose, G. (2001) *Visual Methodologies: An Introduction to the Interpretation of Visual Materials*. London: Sage.
This textbook covers a variety of methodological approaches in visual studies stressing how they may complement each other. A particularly good starting point for students and scholars considering undertaking studies of visual material.

References

Ateljevic, I., Harris, C., Wilson, E., and Collins, F. L . (2005) 'Getting "entangled": reflexivity and the "critical turn" in tourism studies', *Tourism Recreation Research* 30: 9–21.

Ball, M. and Smith, G. (2001) 'Technologies of realism? Ethnographic uses of photography and film', in P. A. Atkinson, S. Dalamont, A. J. Coffey and J. Lofland (eds) *Handbook of Ethnography*. London: Sage.

Beck, U. and Beck-Gernsheim, E. (1995) *The Normal Chaos of Love*. Cambridge: Polity.

Barthes, R. (2000) *Camera Lucida*. London: Vintage.

Barthes, R. (1977) *Image-Music-Text* [essays selected and translated by Stephen Heath]. London: Fontana Press.

Bell, P. (2001) 'Content analysis of visual images', in T. V. Leeuwen and C. Jewit (eds) *Handbook of Visual Analysis*. London: Sage.

Bærenholdt, J., Haldrup, M., Larsen, J. and Urry, J. (2004) *Performing Tourist Places*. Aldershot: Ashgate.

Chambers, D. (2001) *Representing the Family*. London: Sage.

Coleman, S. and Crang, M. (eds) (2002) *Tourism: Between Place and Performance*. Oxford: Berghahn Books.

Crang, M. (1999) 'Knowing, tourism and practices of vision', in D. Crouch (ed.) *Leisure/Tourism Geographies: Practices and Geographical Knowledge*. London: Routledge.

Crouch, D. (2003) 'Spacing, performing, and becoming: tangles in the mundane', *Environment and Planning A*, 35: 1945–60.

Edwards, E. (1999) 'Photographs as objects of memory', in M. Kwint, C. Breward and J. Aynsley (eds) *Material Memories: Design and Evocation*. Oxford: Berg.

Edwards, E. (2002) 'Material beings: objecthood and ethnographic photographs' *Visual Studies*, 17 (1): 67–75.

Edwards, E. and Hart, J. (2004) 'Introduction: photographs as objects', in E. Edwards (ed.) *Photographs Objects Histories: On the Materiality of Images*. London: Routledge.

Feighey, W. (2006) 'Reflexivity and tourism research: telling an (other) story', *Current Issues in Tourism*, 9: 269–82.

Giddens, A. (1992) *The Transformation of Intimacy*. Cambridge: Polity.

Goffman, E. (1959) *The Presentation of Self in Everyday Life*. New York: Anchor Books.

Haldrup, M. and Larsen, J. (2003) 'The family gaze', *Tourist Studies*, 3 (1): 23–46.

Haldrup, M. and Larsen, J. (2010) *Tourism, Performance and the Everyday*. London: Routledge.

Hirsch, M. (1997) *Family Frames: Photography, Narrative and Postmemory*. Cambridge: Harvard University Press.

Holland, P. (2001) 'Personal photography and popular photography', in L. Wells (ed.) *Photography: A Critical Introduction*. London: Routledge.

Hurdley, R. (2007) 'Focal points: framing material culture and visual data', *Qualitative Research*, 7: 355–74.

Krippendorf, K. (2004; 1st ed 1980) *Content Analysis*. London: Sage.

Kuhn, A. (1995) *Family Secrets: Acts of Memory and Imagination*. London: Verso.

Larsen, J. (2005) 'Families seen photographing: the performativity of tourist photography', *Space and Culture*, 8 (3): 416–34.

Larsen, J. (2008) 'Practices and flows of digital photography: an ethnographic framework', *Mobilities*, 3: 141–60.

Latham, A. and McCormack, D. (2009) 'Thinking with images in non-representational cities: vignettes from Berlin', *Area*, 41: 252–62.

Lister, M. and Wells, L. (2001) 'Seeing beyond belief: cultural studies as an approach to analysing the visual', in T. V. Leeuwen and C. Jewitt (eds) *Handbook of Visual Analysis*. London: Sage.

Lorimer, H. (2005) 'Cultural geography: the busyness of being "more-than-representational"', *Progress in Human Geography*, 29: 83–94.

Lutz, C. and Collins, J. (1993) *Reading National Geographic*. Chicago, IL: University of Chicago Press.

Miller, D. (ed.) (2001) *Home Possessions: Material Culture behind Closed Doors*. Oxford: Berg.

Morgan, N. and Pritchard, A. (2005) 'On souvenirs and metonymy: narratives of memory, metaphor and materiality', *Tourist Studies* 5: 29–53.

Pink, S. (2003) 'Interdisciplinary agendas in visual research: re-situating visual anthropology', *Visual Studies*, 18: 179–92.

Rose, G. (2001) *Visual Methodologies: An Introduction to the Interpretation of Visual Materials*. London: Sage.

Rose, G. (2003) 'Family photographs and domestic spacings: a case study'. *Transactions of the Institute of British Geographers*, 28: 5–18.

Slater, D. (1998) 'Analysing cultural objects: content analysis and semiotics', in C. Seale (ed.) *Researching Society and Culture*. London: Sage.

Smith, L. (1998) *The Politics of Focus: Women, Children, and Nineteenth-Century Photography*. London: St. Martin's Press.

Spence, J. and Holland, P. (1991) (eds) *Family Snaps: the Meanings of Domestic Photography*. London: Virago.
Thrift, N. (2007) *Non-representational Theories*. London: Routledge.
Urry, J. (1990) *The Tourist Gaze*. London: Sage.
Urry, J. (1995) *Consuming Places*. London: Sage.
Urry, J. (2007) *Mobilities*. Cambridge: Polity.
van Leeuwen, T. and Jewitt, C. (2001) *Handbook of Visual Analysis*. London: Sage.

10 Beyond content

Thematic, discourse-centred qualitative methods for analysing visual data

Joy Sather-Wagstaff

Introduction

Considering that tourist sites are constituted by landscapes, built environments, shopping areas, exhibits, performances and many more visual, observable and behavioural phenomena, the use of the visual as data for analysis is clearly valuable for a number of different tourism research interests, projects, and desired outputs. Qualitative data analysis (QDA) of visual data can be employed to understand visitor responses to interactive exhibits in a historical museum, identify patterns of sociality and shared public space use at a beachside resort, or to assess the importance of authenticity expected by tourists at a heritage site. Such findings can be utilised for development, management, improvement, and best practices in the tourism and leisure fields. As a means to demonstrate the value of QDA of visual data for tourism research, this chapter presents a basic overview of QDA methods for visual data through the presentation of a case study from the author's research on tourism and tourists' experiences at commemorative historical sites. First, the entry points for research projects and the generation of research questions are discussed as a means to identify which of a wide variety of QDA methods is most appropriate for a given project. Second, a specific mode of QDA, discourse-centred thematic analysis, is demonstrated using data from the case study and the applied value of the analysis results is presented. A brief comparison of manual and computer-assisted data analysis methods concludes the chapter.

The case study included here utilises data collected as part of a larger research project on tourism and tourists at the former site of the World Trade Center (WTC) in New York City, 2002–07. This research focused on how tragic events are processually memorialised through the touristic production, construction, performance, and consumption of sites for 'a heritage that hurts' (Schofield *et al.* 2002: 1). Commemorative historical sites to tragic events are not automatically socially important simply because an event occurred; they are instead spaces that are continuously negotiated, constructed, and reconstructed into meaningful places through ongoing human action including tourist visitation. While usually understood as static places of 'official' cultural expression (history and memory), particularly when fully formalised with museums, monuments and memorial landscapes, they are actually sites that both generate and are informed by what Bodnar (1992: 75) defines as 'public' and 'vernacular' memory and history through the activities of tourists and locals. My research on the WTC memorial sites incorporates multiple types of data including photography and site mapping as forms of fieldnotes documenting observational, spatial, and material culture/artefact phenomena. These data are complemented with information on tourists' subjective experiences from conversations and semi-structured interviews along with historical information collection, all of which contribute to, frame, and contextualise the visual behavioural and artefact data analysis and results. The results indicate the value of qualitative visual data research and QDA for generating rich understandings of visitor behaviour and the practical application of this knowledge in the tourism and leisure profession.

Modes of QDA for visual data and points of entry for research

The works of Banks (2001) and Pink (2006), along with other scholars of visual media, argue for the wide-ranging value of the visual for dynamic research on the human experience. Most works focus on images, be they photographs, film, advertising media, art, or postcards, while Emmison and Smith (2000) artfully add in the 'seen' beyond these material visual images, including built environments and their uses, such as shopping malls, and objects or artefacts located in landscapes such as cemetery markers and graffiti. Others include drawings, maps, and visual hypermedia (see Pink, Kürti and Afonso 2004), even more broadly extending our definitions of what constitutes 'the visual' as both research subjects and tools. However, most published works on the visual, including those in tourism studies, focus primarily on visual media itself (the objects for analysis or illustrative representation) and the research results, leaving a lacuna in terms of explicitly discussing the analytical methods used to derive the results. Exceptions to this include Rose's 2007 *Visual Methodologies: An Introduction to the Interpretation of Visual Materials* and van Leeuwen and Jewitt's 2001 edited volume, *Handbook of Visual Analysis* (see the Further Reading section at the end of this chapter). Rose (2007) and the contributors in van Leeuwen and Jewitt (2001) present the numerous qualitative methods utilised for the analysis of visual media including content analysis, psychoanalysis, semiotic/semiological analysis, structural analysis, compositional and aesthetic analysis, audience studies, historical analysis, and ethnomethodology. Some methods are more useful for specific media (e.g., content analysis for photographs, aesthetic analysis for paintings, audience studies for television advertisements) and this, along with the theoretical positioning of the researcher and the research questions posed, can influence what QDA mode or modes are employed in analysis.

The case study presented in this chapter demonstrates one particular QDA method, a discourse-centred thematic mode of analysis, that integrates both visual and discursive data, combines modes used for visual and non-visual data, and is applied to a tourism research project. It builds upon and makes explicitly qualitative and multi-media one of the more popular forms of QDA for visual data, content analysis. While content analysis of visual data can be qualitative, it is most typically a quantitative, numerical analysis of non-numerical, qualitative data such as images in travel advertisements or postcards. As with all QDA, the categories used to code and analyse the qualitative data may derive from existing theoretical frameworks, be emergent from the data through the analysis process, or combine these two approaches, particularly if only a small portion of the data fits into pre-existing categories. Content analysis usually employs only one type of data source (e.g., travel brochures) rather than multiple sources (travel brochures along with interviews with consumers of brochures) but when it does so, a much more context-rich body of results can be generated. Despite the possibilities for context-richness in this approach, the end results are normatively numerical, focusing on frequencies of phenomena and the compartmentalising of social phenomena. However, this is not to say that it does not have value for many research projects. In addition, given that content analysis derives largely from a positivist approach to research, most evident in its focus on rendering a qualitative data set into quantitative results and a preference for pre-existing categories for analysis, it is a surface reading of the data that may only offer us little more than that which is plainly evident to the researcher (Albers and James 1988).

In general, any QDA method can be deductive, inductive, or a combination of the two for the qualitative analysis of qualitative data. The approach presented here is primarily inductive, generating themes, propositions and explanation through the process of data collection and analysis in the context of building upon existing theories and contributing to applied practices. Methods for thematic QDA derive from and are related to what is commonly called 'grounded theory', an approach to analysing qualitative data first proposed by Glaser and Strauss (1967) that has undergone various permutations and changes since the 1960s through use by researchers across multiple social sciences. At its most basic, this approach can be defined as one that employs a rigorous (rather than anecdotal), inductive/emergent, interpretive, non-sequential, flexible framework for analysing qualitative data. This type of inductive, interpretive approach in tourism research is most commonly associated with ethnographic or sociological research on the subjective, experiential nature of travel. The results of such research can, however, be extended to generating practical applications in the development and effective management of tourist destinations. For example, heritage and historical sites whose primary focus is on the quality of visitors' multiple experiences and effective communication of the historical importance of the site can benefit structurally and organisationally from knowledge about tourists' subjective experiences and behaviours at such sites or related locations (Daengbuppha *et al.* 2006).

There are, however, several issues regarding this process in all of its variant practices that require addressing. A first issue is one of entry to research: how does one identify research questions, particularly when not performing purely deductive analysis where research questions and categories for analysis are guided by a pre-determined theoretical framework and the results support or refute an existing explanation? First, while grounded theory proper proposes that we can go into a research project clear of any influence by existing works, this is, in practice, simply impossible as this requires pure objectivity. However, we can aspire to be more objective in research and this is possible through rigorous QDA. Between theory-driven and theory-generating analysis there exists a continuum of practices that allow for quite a bit of mixing of the two while still performing thematic QDA, even if this is seen as somewhat unorthodox (see Boyatzis 1998). The approach presented here takes this middle ground, emerging from within existing work and aiming to expand and build upon it through inductive analysis. In order for the researcher to determine what types of data are to be collected and what analysis method will be employed, initial research questions, an identification of gaps in existing knowledge, and an issue to be researched more in-depth or by using different methods should be identified.

For the project discussed here as an example, the point of entry for generating research questions grew out of an evaluation of the literature on the social production, construction, consumption and performance of place and space (see Lefebvre 1991 [1974] and Low 2000) in the context of understanding tourist practices and tourism as a part of site meaning-making process (see Coleman and Crang 2002 and Edensor 1998). My research questions were formed in response to a gap I found in existing works on commemorative historical sites. The most commonly investigated aspect of commemorative heritage, and historical sites is the social production of places, the 'historical emergence and political/economic formation' of a specific material environment (Low 2000: 127–8). Such works focus on the process of planning, developing, and physically constructing a formal commemorative place, be it a memorial landscape, the installation of a monument, or the establishment of a museum (see Foote 2001 [1997] and Linenthal 1995 and 2001). Participants in the social production process are typically various civil associations, secular or religious institutions, local and/or national government, victims' family organisations, curators and heritage management professionals, and corporate entities. While attention has been given to the social construction and consumption of place, the process of imbuing meanings to a site through individuals' use, memories, and images of that environment (Low 2000) has not been done specifically in terms of ethnographic research on tourists at sites memorialising tragedies of scale.

When acknowledged, tourists are at best represented as passive spectators of these sites and at worst, destructive consumers whose motivations for visiting such sites are highly questionable. This perspective positions tourists to sites of tragic events and places that memorialise such events as curiosity seekers, passively 'gazing at someone else's tragedy' (Cole 1999: 114) at just another stop on the tourist itinerary. Visitors to such sites who are not surviving 'victims or relatives of victims' are categorised as '"casual" dark tourists' (Lennon and Foley 2000: 169) and their experiences are thus considered to be of less social consequence or importance. This is due largely to the fact that the research in these works is primarily observational and does not investigate the tourist experience from the perspective of tourists themselves. Yet attending to the social construction and consumption of such sites by tourists may engender a rich understanding of the multiple ways a visiting public participates in the creation of a complex matrix of place meanings and importance that is processually constructed over time. Thus the following research questions (RQ) can be posed:

RQ1: What social practices are tourists enacting at the site?
RQ2: What generates these practices and do they change over time?
RQ3: How do these practices impact tourists' experiences at the site?
RQ4: What are the social consequences of these acts and experiences?

Existing theory and research was thus used here to generate a novel project and research questions but not to set forth pre-existing categories for analysis, something more common to strict content analysis or deductive thematic analysis.

In order to get at a deeper level of meaning(s) beyond that of basic content analysis, a discourse-centred, qualitative, thematic, semiotic and interpretive mode of QDA is employed. The use of the word 'discourse' to describe this type of analysis requires explanation as it is not limited to just that of 'talk', one of several definitions for 'discourse.' In terms of literal communication, discourses are 'language-in-use or stretches of language (like conversations or stories)' or other oral narrative forms (Gee 1999: 17) as well as anything that communicates cultural information via interpretation, including material culture such as photographs. We also have discourse that 'exists in the abstract as a coordinated pattern of words, deeds, values, beliefs, symbols, tools, objects, times, and places' (ibid.: 19) that, constitutes human culture as a web of meaning and meaning-making. In addition, all discourse is intertextual and interpretive; talk about one thing is always embedded in the context(s) of that which has been said before, contexts are multiple and thus interpretations can be multiple, and to understand any communicative utterance requires interpretive competence. A discursive analysis approach is one that addresses several issues with QDA of visual media such as photographs. The use of photographs as 'evidence' has been critiqued as positivist and fieldwork photographs themselves are indeed often relegated to data for use in a hypothesis-driven, quantitative 'scientific mode' of research (Banks 2001: 171;

Harper 2003) rather than as data for qualitative research. In addition, as with postcards, analyses of photographs often tend towards that of compositional analysis of content, form, angle, depth, framing, and other aesthetic criteria or basic content in terms of what the images 'contain' as empirical evidence and the numerical frequencies of such content. In contrast, a discursive, thematic QDA considers visual media to reveal multiple, deeper, symbolic meanings through cross-analysis with the talk (discourses) and cultural contexts regarding the data represented through the visual. This case study's particular cross-analysis is performed in order to include the experiences of others (tourists) besides the researcher (the author) in terms of engagement with the visual material culture of the commemorative landscape.

Documentary photographs, along with site drawings/maps, participant observation, interview and conversation data, form an ideal data foundation for a discursive QDA as well as a mediation between emic (tourist/subject) and etic (researcher) perspectives, biases and interpretations. In my research I produced a photographic record of the WTC (2002–07) consisting of over 4,300 digital photographs and which constitutes one set of ethnographic fieldnotes (Sather-Wagstaff 2008). Many of these photographs document visitors' activities while engaging with the diverse features of this site. Others document the site and its contents as it changed over time, generating a 'visual notebook' containing an array of data from artefacts, performances, and protests to vendor activities in the areas at and around the WTC. The spaces that constituted key informal, semi-formal and formal commemorative loci at the site were a central focus for documentation. These loci changed year-to-year and sometimes even month-to-month or week-to-week and the photographic documentation of these spatial shifts has proved valuable in tracking and theorising the fluidity and persistence of informal and semi-formal commemorative spaces and practices. Photographs serve here as a means to 'capture different cultural relations to [a processual] temporality' where the past is rendered present 'in order to gain access to the deep sense of events' (Canal 2004: 38). To complement the photographs, I sketched maps with notes and drawings that tracked visitor movement at the sites, vendor activities and goods at and around the areas, and multiple events including performances and protests. Despite the vast number of photographs I produced, they captured only brief moments in time and are thus acknowledged as very partial, incomplete documents of tourists' activities and experiences. They also reveal little or nothing regarding the subjective identities, what visitors were thinking and feeling, or the individual contexts of knowledge that shaped their interpretive experiences at the site. In order to generate some of this missing data, I engaged in informal conversational interactions or semi-structured interviews with tourists at and around the site regarding their experiences and backgrounds.

The basics of thematic, discourse-centred QDA

Thematic, discourse-centred QDA methods normatively used for text-based data such as interview and focus group results, survey results, fieldnotes and print media, can be applied to visual media. The version presented here through the WTC case study is an amalgamation of several different variations commonly utilised in the social sciences. The rationale for this composite representation is that QDA should ideally be flexible and customisable to a given research project rather than follow strict protocols. Most variations of this type of QDA include several core components, albeit in differing forms, and according to Charmaz and Mitchell (2001: 160), these components are:

1 simultaneous data collection and analysis;
2 pursuit of emergent themes through early data analysis;
3 discovery of basic social processes within the data;
4 inductive construction of abstract categories that explain and synthesise these processes;
5 integration of categories into a theoretical [explanatory] framework that specifies causes, conditions and consequences of these process(es).

The precise terms used to describe particular steps or parts in the QDA process as outlined above varies in the literature (see Boyatzis and Charmaz and Mitchell in the Further Reading section) thus setting out general definitions for use while explaining the basic process here is important. First, a 'code' is a label, description, category or definition for a piece (or group) of data and these can be single words, short phrases or sentences. Coding qualitative data is neither as mechanistic nor as simple as the use of the word 'code' may imply. Coding is a process by which you repeatedly examine and review data, in the case study here, photographs, maps and accompanying fieldnotes, and assign codes for these data. Initial coding, also called open coding, is a process of looking at the data multiple times to deconstruct the data into fragments be it by keywords, phrases, similarities, dissimilarities or terse descriptions. Coding can begin as soon as any data is collected; as new data are collected, existing coded data are revisited for the generation of possible additional codes and comparisons.

While traditionally applied to the analysis of text and narratives generated through sociological or ethnographic interventions, visual media such as photographs can also be coded. When doing initial coding of photographs or other visual media of a given site or otherwise visually evident event/behaviours at a site, there are four foundational categories for code identification one may utilise. These categories elaborate on Albers and James' suggestions of 'subject . . . dress . . . presentation . . . [and] surroundings' as worthy categories in photograph analysis (1988: 145):

1 information about the time and place where the photograph was taken;
2 people in the photographic frame (who they may be, ages, dress, etc.);

3 objects (buildings or other structures, natural landscape elements, decorative artefacts, signage, cars or other mobility technologies, etc.);
4 activities and environmental information (what people are doing or not doing, the weather, etc.).

A helpful framework for coding is that of a semiological analysis, particularly that deriving from the works of Peirce (1940) due to its applicability to both verbal and non-verbal sign (representation) categories (see Chandler in the Further Reading section). One of Peirce's triadic models of signs as falling into three possible categories, iconic signs, indexical signs and/or symbolic signs, is particularly applicable for visual media coding in QDA as well as the coding of text transcribed from oral exchanges. Iconic signs have a direct resemblance to that which they represent and thus they compose a realm of codes that are literally evident (e.g., the souvenir stand and contents in Figure 10.2) through our culturally understood notions about photographs replicating a material reality at a particular moment in time. Iconic codes from photographs are also most amenable to basic content analyses for a site in that they can be used to determine frequencies of selected phenomena.

For example, the two photographs on the facing page (Figures 10.1 and 10.2) are two of 860 randomly chosen fieldwork photographs analysed for the WTC research project. The initial coding of this photograph sample derives from understanding photographs as iconic signs. The items appearing in the photographs become codes as do the locations and dates when the phenomena were recorded photographically. As an example of the initial codes generated, the following codes were derived for the visitor-generated informal commemorative assemblage in Figure 10.1:

- careful arrangement
- civilian clothing
- fire ladder/truck toy
- flowers (formal/florist arranged/wrapped)
- material culture assemblage
- police car toy
- popular culture
- public memorial
- religious pamphlet
- robot toy
- sidewalk
- WTC fence /'The Wall', Church St.
- 11 September 2006.

And the following for Figure 10.2:

- commemorative books
- caps/hats (CIA, FBI, FDNY, NYPD, NYC, 9/11, 'I♥NY')

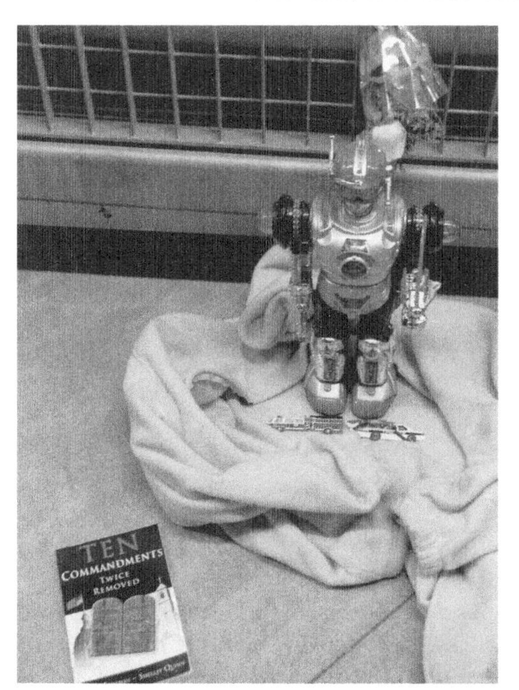

Figure 10.1 Commemorative assemblage detail (2006)

Source: Author's own photograph, used with permission

Figure 10.2 Souvenir vendor table (2003)

Source: Author's own photograph, used with permission

- Church St. @ Liberty St.
- consumer goods
- crystal ornaments/paperweights (Manhattan skyline, angels, WTC)
- T-shirts (FDNY, 'Never Forget', 9/11, 'I♥NY')
- hand-drawn signs
- ice cream truck (food)
- inexpensive ($5 items)
- 14 July 14 2003
- key chains
- lanyards
- Liberty Plaza
- photographs
- portable (tables)
- souvenirs
- storage bins
- street vendor, legal
- US flags.

This type of code is only one subset of coding possibilities and in the case of photographs, initial codes such as these can be used for basic content analysis. Content analyses of visual media can generate quantitative and/or qualitative information such as patterns of movement, repetitions of particular behaviours, or frequencies of specific visual representations. In contrast, discursive, thematic QDA methods produce context-rich information on socio-historical phenomena such as symbolic actions and social interactions, the meanings and values attached to and produced by such, and the social construction of travel itself, tourists' experiences and tourist sites. In order to generate a deeper qualitative analysis that generates identifiable qualitative themes and patterns, one also codes for information beyond that which is literal/empirically present: the symbolic meanings that lie deeper within that which an image represents or, in the case of discourse, talk itself. This directs us to Peirce's two other sign types: indices and symbols.

Indexical and symbolic signs are more arbitrary and lend themselves to a much deeper interpretive schema mediated by the interpretive competence of both tourists and researchers. Indexical signs have an indicative relationship to that which they represent that is culturally understood to be causal even if fundamentally arbitrary (e.g., a teddy bear indexes comfort in many cultures). Symbols are completely contextual and interpretation-based with no 'natural', causal connections between the sign and that which it represents (e.g., the colour black is, in some societies, culturally constructed and understood to symbolise death and mourning while in others it may be the colour white.) While all signs are fundamentally symbols, some are icons, some indices, some both icons and indices. For example, a miniature replica of the WTC twin towers is an icon (it literally resembles the past structure), an index (in one interpretation it points to the events of 11 September as a whole), and a symbol (in one interpretation it represents international power, in another, a loss of power.) Such semiotics-based analyses are frequently used in QDA for tourism research at differing levels of analytic rigour (see Albers and James 1988; Echtner 1999; MacCannell 1999 [1976]; Metro-Roland 2009; Rose 2007 and Urry 2002 [1990] for examples).

An analysis of the data as indexical and symbolic signs yielded a quite different-looking set of codes. A small selection of these deep codes (some quite truncated) that apply to Figures 10.1 and 10.2 as well as to the other photographs from the sample, site drawing and map contents, and transcriptions from conversations with visitors included:

- authenticity
- consumption = commemoration
- enforcement
- force/power
- heroism
- honour
- importance of lost landscape icons (the WTC towers)
- incidental display/participation
- informality
- intentional display/participation
- Judeo-Christian ideologies and symbols
- legality = legitimacy
- marking presence in place
- marking the place of the dead
- materiality of memory
- mourning
- participatory commemoration
- power/force
- protectors/protection
- punishment
- remembrance
- rescue work(ers) as heroes
- revenge/avenge
- superhero(ism)
- symbolic power of the dead
- temporariness
- trust in authority
- truth-seeking.

Discourses on the site and site contents were key to understanding more than just the site itself through photographic data. As a researcher, I documented the site photographically and through drawings and maps while the visitors at the site were engaging with, making, looking at, photographing, and experiencing the site. Data from conversations with tourists at the site were thus critical to both generating deep codes and controlling for researcher bias as the sole interpretant of the material culture and social phenomena of the site through field photographs. The initial and deep codes were generated as the visual and discursive data was deconstructed and these were repeatedly read across each other to discover the multiple social processes at work and to identify emergent themes, patterns or clusters that would be quite impossible to see by simply viewing/reading data a single time. Themes (also called axial codes by some) were then generated through a reconstruction of the fragmented data by grouping or linking data; these groups or links are sometimes directly and easily observable through the codes or they may reveal 'hidden' structures underlying the codes. Some key high-order social process themes relevant to the discussion here that emerged from QDA performed on the WTC photographic data include:

1 participation in public memory-making and performing grief rituals;
2 persistence in creating memorial assemblages as a means of marking presence;
3 resistance to rules/limitations on both informal commemoration and commerce;
4 importance of participation as part of tourists' most meaningful experiences;
5 consumption and commerce as necessary to commemorative practices.

These themes were further reconstructed into even more discrete, meta-level categories. Meta-level categories for the photographs from the WTC site include: *performance*, *consumption*, *materiality*, and *participation*. These categories were used to build explanatory frameworks for the 'specific causes, conditions and consequences of these [social] process(es)' (Charmaz and Mitchell 2001: 160). These frameworks can be used to generate suggestions for best practices applicable to future commemorative historical landscape or museum sites that will, undoubtedly, be destinations for both tourists and local visitors.

As an example, I present four interrelated suggestions for consideration in planning, developing and maintaining such sites. All of these suggestions are grounded in an understanding of tourists as more than just consumers of sites made for them; they are participants in the social construction of sites as invested agents in meaning-making and this is a position developed through the identification of the five high-order social process themes listed in the previous paragraph. The suggestions derive from identifying and understanding the social and individual consequences of engaging in consumptive, performative and constructive activities: symbolic interaction with the dead, comfort and catharsis, participation in making the site historically and emotionally salient both individually and collectively, and the embodied production of memories and historical knowledge that when shared with others, constitute the site as important for those who may visit in the future. First, recognise the importance of allowing for or even pre-establishing official spaces for informal, participatory commemorative expression by both tourists and visitors such as that represented by the commemorative assemblage in Figure 10.1. Second, such visitor-generated commemorative assemblage spaces must be managed and this impacts planning for site staffing and volunteer worker numbers and training on site maintenance. Third, should a decision be made that assemblage items be collected and archived, space and staff must also be allocated for such and decisions made regarding what items will be collected versus those that will be removed and destroyed. Fourth, given that souvenir items play a role in both commemorative assemblages and as the material culture of memory for tourists, careful consideration must be taken regarding the issue of whether commerce is allowed or not and if so, where it is allowed and what types of items are appropriate for sale.

Computer-assisted QDA

While the discourse-centred, thematic QDA process for the case study presented here was performed mainly through manual labour combined with the use of some software, there are numerous software suites worth mentioning that are used to perform various modes of QDA. Among the most popular suites in use across academic disciplines, government agencies, research and development programmes, and consulting businesses are ATLAS.ti, NVivo (formerly NUD*IST, Non-numerical Unstructured Data Indexing Searching and Theorising) and HyperRESEARCH, as they can integrate multiple media types including video, graphics, PDFs, and audio files in addition to text and numerical data. Some QDA software has the capability to generate statistical data for analysis (a most basic content analysis form) in addition to purely qualitative themes and theoretical categories. In terms of tourism research, computer-assisted analyses have traditionally been limited to basic numerical or complex statistical marketing use-oriented research whereas it can also be used in qualitative research on tourists' phenomenological experiences for multiple applications. Mehmetoglu and Dann (2003: 7–8) provide a thorough overview of ATLAS.ti specifically for tourism research of this kind, while Fielding (2001) presents ETHNOGRAPH and HyperRESEARCH, with both works providing a comparison of traditional and computer-assisted QDA. These authors note that the main strengths of computer assistance are:

1. time savings in terms of hours spent performing analysis;
2. reductions in multiple types of human errors, particularly when analysing very large data sets;
3. the possibility of a more inductive analysis of the data in contrast to the imposition of pre-existing, deductive frameworks on the data by researchers.

While these are indeed significant benefits, one must also consider the disadvantages to using such software, both technical and in terms of analysis results.

First, the computer operating system (Windows/Vista, UNIX or Macintosh) that one utilises will significantly limit QDA software options. Mac users are at a clear disadvantage when it comes to computer-assisted QDA as the most popular and broadly used programmes will not run on a native Mac operating system (MacOS) and a few only sometimes successfully run with the use of Boot Camp or other PC-mirror software installed. For MacOS, a highly-rated QDA software suite is ResearchWare's multimedia HyperRESEARCH and HyperTRANSCRIBE bundle and there is a free, open-source, text media-only TAMS (Text Analysis Markup System) Analyser for MacOSX. A second critical issue with software is cost. Freeware such as TAMS tend to be bare bones and may only be in beta form while multimedia analysis software start at around €1,000 for a single-user licence. Most companies do offer significant discounts for educational institutions and multiple-user licences but for graduate students or researchers without technical support funds, costs may still be prohibitive. However, in many cases, investment in the software may mitigate the numerical costs of labour hours spent performing manual analyses.

A third issue with using QDA software is unrelated to costs or operating system specifications and directly related to a fundamental tenet of QDA: a true closeness with the data. Researchers' visual and tactile intimacy with the data is viewed as key to success; if this is removed from the process, it could lead to a mere surface analysis (see Fielding 2001). When performing a manual analysis, the researcher spends considerable time going through the data repetitively by hand in order to generate the codes, emergent themes and patterns. The level of familiarity the researcher develops through this process allows for a deep feel of the data and the possibility of identifying underlying patterns and themes that are quite innovative and creative. As Mehmetoglu and Dann (2003: 7–8) note, hands-on manual analysis also precludes linguistics-based errors given that computer analysis may incorrectly identify and code certain individual parts of speech that are context-specific in terms of intended meanings. Most software suites allow for manual corrections to these errors or entry of manual instructions for automated correction. Fourth, there remains a certain amount of manual labour required when using software for analysis. If a transcription programme that can transform recorded audio data or handwritten notes is not used, manual transcription of such is required. Likewise, any raw data must be put into forms that can be uploaded into and read by the software. Finally, the data must be divided into relevant units by the researcher for loading and often initial coding must be performed manually before proceeding with the computer-assisted analysis to identify code groups, families, connections, and themes/patterns. Computer-assisted QDA is thus not a fully automated, self-contained process but one that may still require considerable manual labour inputs from researchers.

Some researchers will mix both manual and computer software-assisted analysis methods and this is one of the best ways to familiarise oneself with the limitations and advantages of each. I rely primarily on manual analysis with some use of software, particularly to cross-check manual results from large data sets. Database software can also be exploited as a means for organising, coding and linking data for the manual analyses of such. While I use TAMS Analyser for some text analysis, I employ iView MediaPro, a database programme for multiple forms of media kept in multiple computer and external drive locations that allows for the manual organisation, coding, and linking of digital photographs, maps, notes, audio files and other collected media. iView allows keywording and this process can be used to record codes. It also allows for manually linking items to one another (not unlike the automated linking in software like ATLAS.ti) thus a photograph or unit of photographs can be linked to a relevant news media PDF, interview transcript, map, or other fieldnote document. The codes and links are searchable and a 'notebook' of codes and links as search terms can be generated. As a final note, it is highly recommended that one learns manual analysis methods prior to employing software for QDA. As with quantitative statistical analysis, learning the hands-on methods first forms a solid foundation for fully exploiting computer-assisted analysis methods while also allowing the researcher to become more attuned to possible human and software errors in coding, and engage in deep interpretation, and to problem-solve.

CHAPTER SUMMARY

- There are multiple modes of QDA that can be employed in the analysis of visual data and each is specific to the theoretical positioning of the researcher, the research project, and often the type of media being analysed.
- Content analyses of visual data generate surface-level knowledge of behavioural phenomena, typically quantitative, while context-rich, thematic QDA can produce deeper-level knowledge.
- Semiotic analysis models can be helpful in identifying deep, interpretive codes during the analysis process.
- QDA, when used to analyse multiple sources of data, enables tourism researchers to produce rich tourists' experience-based explanatory frameworks that go beyond mere description and which lend themselves to generating practical applications based upon the knowledge produced.
- Qualitative data, including that from visual media, can be rigorously coded and analysed manually, through the use of data analysis software suites, or with a combination of manual and computer-assisted methods; each of these methods has its advantages and disadvantages.

Annotated further reading

Boyatzis, R. (1998) *Transforming Qualitative Data: Thematic Analysis and Code Development*. London: Sage.
Boyatzis bridges the often separated worlds of quantitative versus qualitative analysis and theory-driven versus inductive analysis through an introduction to basic thematic analysis. The text provides step-by-step instruction on coding, developing themes and other units of analysis, sampling, and determining reliability while also addressing the practical and theoretical challenges of doing thematic analyses of qualitative data.

Chandler, D. (2007 [2002]) *Semiotics: The Basics*. London: Routledge.
Chandler provides an expansive, hands-on, example-driven explanation of the basics of semiotic theories, modes of analyses, and the origins of semiology. Topics include structuralist and post-structuralist theories, sign models, codes, rhetorical modes, and textual interaction as applied to text, talk, mythology, film, photographs and many other elements of culture and society.

Charmaz, K. and Mitchell, R. G. (2001) 'Grounded theory in ethnography', in P. Atkinson, A. Coffey, S. Delamont, J. Lofland and L. Lofland (eds) *Handbook of Ethnography*. London: Sage.
Charmaz and Mitchell introduce the origins of grounded theory and the various permutations it takes for the qualitative analysis of qualitative data. A foundational framework for the process of grounded theory based QDA (including coding, sampling and writing issues) and interpretive, qualitative analysis is presented step-by-step and illustrated with examples from case studies.

Rose, G. (2007) *Visual Methodologies: An Introduction to the Interpretation of Visual Materials*, 2nd edn. London: Sage.
Rose's text is a highly organised and accessible presentation of a spectrum of methods (including content analysis, psychoanalysis, and semiotic analysis) for analysing a range of visual data from film, art and photographs to places and television; specific terminology is clearly defined, the research process is discussed in detail, and recommendations for further reading are made.

van Leeuwen, T. and C. Jewitt (eds) (2001) *Handbook of Visual Analysis*. London: Sage.
Van Leeuwen, Jewitt and their contributors from linguistics, cultural studies, anthropology, literature, and film studies provide a series of case studies in visual data analysis that span magazine and fieldwork photography to geography textbook photographs and children's drawings. Methods of analysis presented include content analysis, social semiotics, ethnomethodology, and psychoanalysis.

References

Albers, P. and W. James (1988) 'Travel photography: a methodological approach', *Annals of Tourism Research*, 15: 134–58.

Banks. M. (2001) *Visual Methods in Social Research*. London: Sage Publications.

Bodnar, J. (1992) *Remaking America: Public Memory, Commemoration, and Patriotism in the Twentieth Century*. Princeton, NJ: Princeton University Press.

Boyatzis, R. (1998) *Transforming Qualitative Data: Thematic Analysis and Code Development*. London: Sage.

Canal, G. O. (2004) 'Photography in the field: word and image in ethnographic research', in S. Pink, L. Kürti and A. I. Afonso (eds) *Working Images: Visual Research and Representation in Ethnography*. London: Routledge.
Chandler, D. (2007 [2002]) *Semiotics: The Basics*. London: Routledge.
Charmaz, K. (2000) 'Grounded theory: objectivist and constructivist methods', in N. Denzin and Y. Lincoln (eds) *Handbook of Qualitative Research*. Thousand Oaks, CA: Sage.
Charmaz, K. and Mitchell, R. G. (2001) 'Grounded theory in ethnography', in P. Atkinson, A. Coffey, S. Delamont, J. Lofland and L. Lofland (eds.) *Handbook of Ethnography*. London: Sage Publications.
Cole, T. (1999) *Selling the Holocaust: From Auschwitz to Schindler. How History is Bought, Packaged, and Sold*. New York, NY: Routledge.
Coleman, S. and M. Crang (2002) 'Grounded tourists, travelling theory', in S. Coleman and M. Crang (eds) *Tourism: Between Place and Performance*. New York, NY: Berghan.
Daengbuppha, J., Hemmington, N. and Wilkes, K. (2006) 'Using grounded theory to model visitor experiences at heritage sites: Methodological and practical issues', *Qualitative Market Research: An International Journal*, 9: 4 367–88.
Echtner, C. M. (1999) 'The semiotic paradigm: implications for tourism research', *Tourism Management*, 20: 45–57.
Edensor, T. (1998) *Tourists at the Taj: Performance and Meaning at a Symbolic Site*. London: Routledge.
Emmison, M. and Smith, P. (2000) *Researching the Visual: Images, Objects, Contexts and Interactions*. London: Sage.
Fielding, N. (2001) 'Computer applications in qualitative research', in P. Atkinson, A. Coffey, S. Delamont, J. Lofland and L. Lofland (eds) *Handbook of Ethnography*. London: Sage Publications.
Foote, K. (2001 [1997]) *Shadowed Ground: America's Landscapes of Violence and Tragedy*. Austin, TX: University of Texas Press.
Gee, J. P. (1999) *An Introduction to Discourse Analysis: Theory and Method*. London: Routledge.
Glaser, B. G. and Strauss, A. L. (1967) *The Discovery of Grounded Theory*. Chicago, IL: Aldine.
Harper, D. (2003) 'Framing photographic ethnography: a case study', *Ethnography*, 4 (2): 241–66.
Lefebvre, H. (1991 [1974]) *The Production of Space*, D. Nicholson-Smith (trans.). Cambridge, MA: Blackwell.
Lennon, J. and M. Foley (2000) *Dark Tourism: The Attraction of Death and Disaster*. London: Continuum.
Linenthal, E. (1995) *Preserving Memory: The Struggle to Create America's Holocaust Museum*. New York, NY: Penguin.
Linenthal, E. (2001) *The Unfinished Bombing: Oklahoma City in American Memory*. New York, NY: Oxford.
Low, S. M. (2000) *On the Plaza: The Politics of Public Space and Culture*. Austin, TX: University of Texas Press.
MacCannell, D. (1999 [1976]) *The Tourist: A New Theory of the Leisure Class*. Berkeley, CA: University of California Press.
Mehmetoglu, M. and Dann, G. M. S. (2003) 'ATLAS.ti and content/semiotic analysis in tourism research', *Tourism Analysis*, 8:1–13.

Metro-Roland, M. (2009) 'Interpreting meaning: an application of Peircian semiotics to tourism', *Tourism Geographies*, 11 (2): 270–79.
Peirce, C. S. (1940) 'Logic as semiotic: the theory of signs', in J. Bucher (ed.) *The Philosophy of C.S. Peirce: Selected Writings.* London: Routledge.
Pink, S., Kürti, L. and Afonso, A. I. (eds) (2004) *Working Images: Visual Research and Representation in Ethnography*. London: Routledge.
Pink, S. (2006) *The Future of Visual Anthropology: Engaging the Senses*. London: Routledge.
Rose, G. (2007) *Visual Methodologies: An Introduction to the Interpretation of Visual Materials*. London: Sage.
Sather-Wagstaff. J. (2008) 'Picturing experience: a tourist-perspective on commemorative historical sites', *Tourist Studies: An International Journal*, 8(1): 77–103.
Schofield, J., Johnson, W. G. and Beck, C. M. (2002) 'Introduction: matériel culture in the modern world', in J. Schofield, W.G. Johnson and C.M. Beck (eds) *Matériel Culture: The Archaeology of Twentieth-century Conflict*. London: Routledge.
Urry, J. (2002 [1990]) *The Tourist Gaze*. London: Sage.
van Leeuwen, T. and C. Jewitt (eds) (2001) *Handbook of Visual Analysis*. London: Sage.

11 Representing visual data in tourism studies publications

Richard Tresidder

Introduction

The use of visual materials within tourism research has a long and academically significant history; however visual data can sometimes be difficult to publish because of a number of constraints including the publishing format, the cost of reproducing visuals, and the difficulties involved in obtaining copyright. However, researchers should not be discouraged from engaging in visual research as it offers the opportunity to connect with many of the discourses that inform contemporary tourism studies. This chapter explores some of the barriers and strategies that may be used in the presentation of visual data in publications and is organised into two sections. Section 1 examines some of the practical considerations that need to underpin the use of visual images within the research process, and considers issues such as the ethics and legal requirements of presenting visual data within publications. Section 2 investigates some of the available techniques, considerations and processes that enable moving and still visual images to be successfully utilised within tourism publications and research.

Sourcing visual materials

As indicated in many of the chapters in this volume, there are many different types of visual data that may be drawn from both primary and secondary sources. There is a significant range of visual materials in the form of secondary data that surround both tourism as an activity and tourism as an industry. For centuries tourists, hosts and academics have recorded the sights and experiences of tourism in many mediums from paintings, to postcards, to photographs and home movies. While the intangibility of the tourism product has required the development of a creative approach to product marketing, consequently, tourism marketing and promotional materials have relied on creating a visual representation of the destination and the types of experience the tourist can expect to consume once there. This has produced a large archive of posters, postcards, brochures, websites, flyers, film-based adverts and promotional logos that are worthy of analysis. Moreover, the significance of tourism as a social and cultural activity has led to destinations, events, festivals, food, tourist behaviour and impacts being charted in travel, reality, home buying, comedy and drama television programmes as well as the use of destinations and attractions in films, each one contributing to the visual discourse of tourism. These are also supported by the primary research of researchers who are generating visual materials in the field and are themselves further contributing to the visual discourse of tourism.

This rich archive of secondary and primary visual materials supports many research approaches and interests; as a result, the way in which researchers utilise and represent visual materials in publications will depend upon the approach and objectives of their research. For example if researchers are looking to identify patterns of usage or themes within tourism research then it is likely that their sample will be quite large. As a consequence, the question is whether it will be possible to use a large number of visuals in the publication or whether only images that act as an exemplar of significance should be used instead. It is important to also establish whether permissions from all sources can be obtained. If researchers are trying to illustrate a process, then it is likely that not as many examples will need to be used within publications. An in-depth semiotic analysis on the other hand may require the analysis of one source. In addition, it might be necessary to determine whether still or moving images will be used in a project. Each of these approaches will have very practical implications for how researchers choose to present their visual data in publications and will also reflect on the size, position, and number of images included.

Section 1: Practical considerations for the use of visual images within tourism research

Tourism studies draw from a wide range of visual materials and whether these materials are produced by the researcher or are sourced from other people or sources such as the Internet, there are a number of practical and legal issues that need to be identified and considered before images are used in the context of tourism research and publications. This section examines the notions of ownership of visual data. First, the use of visual materials in research must be balanced against certain legal, philosophical and moral considerations. The materials utilised will often have been produced by a third party such as, for example, images in tourism brochures or postcards, and the ownership of those images may be with an individual or an organisation. As a result, the use and presentation of those images within academic publications will require proper permissions to be given by the owner. Second, even if researchers produce their own images as part of the research process, the use of images of people or material forms of culture, such as for example pictures taken of tourists or those attending festivals, will also raise ethical, moral and governance issues. Such methodological approaches also raise questions as to the rights of those represented in the image to be the focus of analysis or to be represented as part of a publication resulting from a research project. As such any research project needs to ensure that both copyright and ethical considerations have been adequately considered and dealt with.

Research ethics and governance

Visual research like all research is underpinned by the guiding ethical principles of non-malfeasance (the principle of doing or permitting no official misconduct) and beneficence (the requirement to serve the interests and well being of others, including respect for their rights). These principles mean that there is a systematic regard for the rights and interests of others within the research process. This is reflected in the regulation of research by a range of legal, governmental and organisational processes that oblige the researcher to behave in an ethical manner (Mertens and Ginsberg 2009). In practice this means that researchers need to ensure that the research they undertake does not impact upon individual's rights (see Wiles *et al.* 2008), the two overriding principles of which are:

- *Autonomy*: People must be free to make their own informed decisions.
- *Justice*: People must be treated equally, and the researchers will need to inform participants about confidentiality, voluntary participation and that their research is independent and avoids causing harm.

Traditionally the use of images representing clearly identifiable people has been the focus of visual research and has often not involved consent (e.g., see Prosser 1998; Banks 2006). However over the past five years the level of ethical scrutiny and regulation in academia has increased to the point that 'regulation will render aspects of visual research virtually impossible, or at best will place severe limitations to visual researchers' practice' (Wiles *et al.* 2010: 1). However, the use of visual data is still particularly significant within tourism whether it is being used for commercial or research purposes as it provides a rich source of information. Importantly, much of this visual data may be used for both research and within the representation of research findings as long as researchers adopt an ethical approach.

Therefore the utilisation of any visual images within tourism research that involves the representation of identifiable 'people' within publications must be treated with caution as these people become 'represented participants' (Kress and van Leeuwen 1996). Therefore, it is the researchers' duty to protect the rights of those photographed or filmed. There are many good texts such as Porter and McKee (2009) that provide guidance through the ethical process of using visual media within research. However, researchers should also check procedures with their university and other relevant academic institutions which have a policy and process which they are expected to follow and which may also provide local guidance. The checklist provided below could be used as the minimum basis for any research process where identifiable individuals are represented:

CHECKLIST 1

1. Have you obtained consent from those represented within your research to use the images for the purposes of research and subsequent publications?
2. Have you informed them of the nature, role and status of your research?
3. Have you the ability to maintain their anonymity/confidentiality if required?
4. Can you be certain that your research will not cause any harm?
5. Have the participants agreed voluntarily to be part of the research?

It is important to note that these are general guidelines, however if the research involves children, medical contexts or vulnerable people you should seek further advice as the ethical considerations for these groups are often more rigorous (see Heath *et al.* 2007).

Ethical strategies for the presentation of visual data

One of the major ethical considerations involves protecting the anonymity of research participants, and this often provides barriers to using images of people and places within visual research publications. However, it is possible to develop a practical approach to the use of visual images that ensures that the rights of individuals are protected (see Wiles *et al.* 2008 for a comprehensive overview). In cases where this is deemed to be the most suitable solution, this might involve obscuring details of people or places in order to ensure the anonymity of the research participant. Practical techniques for obscuring the identity of both people and places include:

- Obscuring identity in images: there are a number of websites and software, such as Photoshop or shareware such as WAX 2.0, that enable you to increase the pixilation of features in order to blur the image (both still and moving), convert images into cartoons or drawn images, and to blocking out distinguishing features.
- Obscuring place: often if individuals are obscured then the significance of place identity is not so significant. However, if your research is focussed in a particular geographical location then a community or group may be identified via their association with that locality. As such, if you feel that

the anonymity of a group is at risk then you could pixilate or hide store or house names or other distinguishing geographical characteristics from the reader.

Although providing anonymity by obscuring places and faces can indeed be a very useful approach in some visual research contexts, not all researchers will necessarily agree or subscribe to this practice in all of their visual research projects. In fact, the 'approach favoured by social researchers is to present visual data in its entirety' (Wiles *et al.* 2008: 27). Namely, if deemed as central to the research project and the representation of research findings, there are a number of modes in which images can be ethically included in publications without obscuring details in the image as long as the ethical or legal rights of the subject are not threatened. These might include using images taken in public places where there are no legal or other restrictions for photography or videomaking, or using images for which an informed consent and/or copyright clearance has been obtained (e.g., see Banks 2006 and Wiles *et al.* 2008 for a further discussion on this). Nonetheless, even in these cases researchers will still need to be cautious and consider any potential issues and contexts in which it would not be ethical to include such images. For example, even if an image has been taken in a public place where no restrictions to photograph or film existed; or if, for example, the people who appear in the image seemed to be doing something private in a public place, it would not be ethical to include the image in a publication. Therefore, in some cases it is a subjective decision that the researcher makes as to the impact of their actions.

Copyright: ownership of materials

With the advent of Internet search engines and online image libraries it has never been easier to access visual data and material for tourism research, however much of the material available is still subject to copyright protection. The copyright of visual materials and sources provides an automatic legal framework that protects materials from unauthorised usage, such as their unauthorised reproduction in academic publications. Therefore it is essential for researchers to gain permission to include any images or material that they wish to include within publications. There are a number of texts that will guide researchers through the process such as Shay (2009) and Stokes (2009). Furthermore, there are also a number of web-based resources such as www.JISC.ac.uk which provide online materials for academics and researchers. Additionally, most academic institutions will also have a specialist within the library or learning centre who can help researchers negotiate the copyright rules and regulations.

Researchers wishing to incorporate visual data in their publications should not let copyright concerns deter them from following visual research as most organisations and companies will allow their images to be used if it is for academic purposes. It is important, however, that researchers follow the guidelines shown below:

CHECKLIST 2

1 Have you identified the image, where it is sourced, what you want it for and where you plan to publish it? This will help you identify what type of permission is required.
2 Are you going to use the image more than once? If you are, is the purpose and context the same as the permission originally given?
3 Have you got copyright permission in writing? This should ideally be either in writing in the context of images sourced from secondary data, or if images are created as a part of a research project, informed consent would need to be obtained in a written or recorded (audio-visual) form.
4 Have you made every possible effort to contact the person or organisation that produced the image? Keep evidence of your attempts and check with a copyright specialist that it is OK for you to use the image.
5 If you do not have permission or informed consent ideally you should not use the image.

Publishing visual research: checking publisher's guidelines

If writing for publication, researchers should first check the publisher's webpage as one of the major barriers apart from copyright issues is the cost of reproduction. As a result, many publications will have a strict limit on the number and type of images that may be used. There is also a growing trend for publishers to charge the author a fee if they wish the images to be in colour rather than black and white, as a consequence this may limit how and where researchers present their visual data. However, by adopting certain strategies it is possible to effectively represent and develop visual research in an effective and interesting manner. These strategies include:

- developing an external website that holds the images to which you can refer the reader;
- developing your descriptive skills so you can describe the nature, aesthetic and context of the image;
- checking that the images identified actually give the work added value.

The first strategy provides an opportunity to present a great deal of data in the manner in which it is desired that it be seen. The second strategy is more about developing writing and descriptive style, a skill that is usually developed through extensive reading of other authors' descriptions of visuals in publications. Travel writing texts could also be examined for tips and style pointers. The role of the researcher as an author is to describe and show the reader the image through words and as such researchers should suspend personal judgements during the descriptive stage and leave this for when an analysis of the description is provided (for good examples see Messaris 1997). One of the methods in achieving this is to allow nouns and verbs to do the work of description. With nouns, the readers will see; with verbs, they will feel. The text has to become the readers' eyes.

The key to identifying and using visual images as an element of the research process is to identify and adopt the most appropriate method or approach and the next section of this chapter considers some of the practical considerations you need to undertake in presenting visual materials within research.

Section 2: Approaches

This section of the chapter will enable visual tourism researchers to develop the method selected to present visual data. The section consists of four elements: using visual images within the interpretivist and positivist dichotomy, using photographs, using moving images and contextualising visual data.

Presenting visual data within the interpretivism vs positivism debate

One of the barriers to the publication of visual research within tourism studies is that it often provides a subjective evaluation of images rather than providing the perceived certainty and validity of positivist research. For example, Barthes (1964) considered photographs to be polysemic and thus capable of generating multiple meanings for both the researcher and the audience. Many tourism publications still require that the research is underpinned by a positivistic and quantitative methodology. However, it may be argued that the analysis of images provides us with an understanding of how people interpret, understand and chart the tourism environment. In other words, visual approaches generally adopt a method and research approach that emphasises meaning and subjectivity set against the instrumental logic of positivism (Botterill *et al.* 2001). The need to recognise the importance of the individual within the interpretation process is therefore central to the use of visual data (Kress and van Leeuwen 1996) in tourism studies.

The choice of visual images, the medium and the subsequent interpretation by the researcher or in certain cases the participant, is an example of the reflexive element of the research and interpretation process (informed by differing social, cultural and geographical knowledge) and is reinforced by certain socially and culturally embedded definitions of tourism. The choice of visual medium may be seen as just one genre of text which locates the researcher and participants' 'being-in-the-world' and will reflect certain shared views or 'consensus constructs' (Botterill *et al.* 2001) of the world.

For example, the individual's analysis of home holiday movies will depend upon their experiences of holidays, their technical knowledge and their relationship to the participants, in other words their 'being in the world'. This is reinforced by Schwartz (1989: 120) who states that 'photographs embody the personal concerns of the photographer-artist'. Thus it is difficult to separate visual methodologies and choices of images, 'represented participants' or subjects from the researcher's own epistemological and ontological foundations. Therefore, the choice of images and the subsequent justification of these images within one's methodology, become an even more significant element of the research process as researchers need to methodologically defend and support the integrity of their chosen approach within both the context of research and subsequent publication of findings.

Photographs

Still photographs as primary and secondary sources generate a rich stream of research data for the researcher within tourism studies, as the intangible experience of purchasing a holiday is informed by images within marketing texts, as tourists chart their experiences through photography, and as academics and researchers generate primary research material by charting activities and experiences through still photographs. However, the use and significance of the research and images chosen depends upon the nature and context of how the pictures are used by the picture makers and the viewers. That is, the image of a beach represented within a tourism brochure is very different from that taken by a tourist, the beach may be the same, but the purpose, context and meaning will differ as the sign vehicle (brochure or photo album) have very different purposes.

In addition, the relationship between the researcher and the reader/viewer of the research is problematic as the researcher has no control over how the reader interprets the photograph. Schwartz (1989) states that viewers generally possess a twofold context-specific definition of photography; the first definition identifies photography as art and raises concerns as to vision or

CHECKLIST 3

Is the photograph's purpose to:

1 Support an assertion? The photograph becomes an 'illustration' of something already described within the text.
2 Confirm presence? The photograph becomes a research document that proves you were there or an event happened.
3 Focus on the research? The photograph becomes the primary means of promoting response or unit of analysis.

motivations of the artist. The second defines photographs as a precise machine-produced record of a scene or subject and how accurate this representation is. Therefore, before utilising and presenting photographs there is a need to identify the context of the image within the research project (e.g., Banks 2006). As a consequence, the first stage of the decision making process is to decide the context of the photograph within research (see, for example, Checklist 3, on p. 194). Thus, the way in which images are presented and utilised within the text varies according to purpose and context of the photograph and research.

If one's research falls into the first two categories then the significance of the images are significantly less than in the third category as the photograph plays a supporting role within the work. The consequence of this is that researchers can be more selective about the number, type and technical production of the image, as the nature of the research (image as support) does not require the same level of photographic detail within the publication of the research findings. However, through adopting the photograph as the means of generating theory or data, the significance of the image becomes more significant and subsequently, needs locating within both a research and presentational methodology. Just as researchers learn to structure arguments and debates and develop a writing style, the same applies to visual methods. Although part of this includes developing technical skills in terms of producing and editing images, the biggest challenge is to sequence material to create a coherent and logical narrative structure. Banks (2006) takes this idea a step further by discussing the use of photography in visual research as developing a 'Photographic Essay'. The idea of this is to develop a reasoned argument in the style of photo-journalism. This is a means of attempting to reduce the subjective experience of the researcher's or participant's knowledge informing the logic and direction of the visual argument. It is argued that the sequencing or order of the photographs should tell a story that the researcher is part of, but should become primarily the story of the research subject or represented participant. The development of this approach is to counteract criticism of such research being overly subjective. However, it may be argued that the interpretation of any photographs is still a subjective interpretive activity.

Three approaches of using photographs as the primary element of the research process are identified by Heisley and Levy (1991) as creating cultural inventories; this would be utilising photographs in tourism to examine a cultural movement, whether that is tourist behaviour, cultural impacts of tourists or events and festivals. Second, as project stimuli; for example this may be used where you are trying to examine tourist's memories of a destination by exploring their family photographs or showing them images of a destination. Finally, where photographs are used to examine social and cultural artefacts, for example in tourism studies this may involve using images of indigenous art or architecture as the unit of analysis. Each of these approaches requires a different means of representation as the purpose of the research and its publication and the use of the image comes from a different context.

Photographs as cultural inventories

Visual images can be used to create a cultural inventory of people, objects and events whereby 'the camera's eye can extend and refine scientific description by including detail and nuance' (Heisley and Levy 1991: 259). In this context that camera becomes an objective recorder recording valid and reliable data. As undertaking a cultural inventory requires a microanalysis and 'microimage' analysis in order to deconstruct or decompose the image, the presentation and the quality of the image need to be of a size and standard that enables this analysis to take place. Thus, the significance of the photograph within this context is particularly high and as such requires a skill and access to technology to support this visual approach.

Photo-elicitation

Photo-elicitation can be defined as a projective method in which we use photos to stimulate a social memory, and which enables the elicitation of an individual's personality and cultural values (Heisley and Levy 1991; Banks 2006) within the research. The significance of the method in terms of presenting the photograph in the research is not so great, as the significance is the individual's response to the images rather than presenting the image itself. Thus, the presentation and production of the image within text becomes a reference point to illustrate the findings rather than being the primary focus of the research. Therefore the photograph plays a supporting role, but still retains a level of significance that requires a presentation approach that sees the image weaved into the work (for a good example of this presentational approach see Schwartz 1989).

Photographs as social artefacts

The presentation of the visual image as an example of a form of social artefact, and where the photograph comes to represent and indicate the values of the participant or of a group, elevates the significance of the photograph within the text as the image becomes the findings of the research and the focus of any analysis. Although you may adopt a microanalysis or microimage analysis of the picture, the presentational approach differs from that of the photograph as cultural inventory and benefits from being presented in the form of a visual essay, creating a logic and structure to the work, yet does not require the same level of technical development as photo-elicitation. Many of the themes and guidelines developed in this chapter can also be applied to moving images.

Moving images

The use of films and moving images within research presents a number of issues, and these are based around the role of the reader as a reflexive being who interprets and reads the images presented to them according to their own social and cultural background and their ontological and epistemological foundations. The result of this is that the showing of films to the reader promotes a reader response and ultimately splits audiences. Research undertaken by Martinez (1990) demonstrated that the more didactic and authoritarian the narration is within the film, the more it is open to aberrant readings, whereas those films that are more open engage the viewer further because they encourage thought and interpretive effort and require greater effort to find meaning. For Martinez, reflective reading of the film by viewers was aided by encouraging viewers to recognise that the ethnographic films are a particular genre, with particular conventions and history, in other words it is the role of the researcher to 'alert viewers to the external narratives surrounding films, rather than assuming unproblematic and automatic transmission of the internal narrative or content.' (Banks 2006: 141).

One of the issues of utilising moving images within a research project is that there is good evidence (Schwartz 1989; Banks 2006) that viewers of moving images particularly in the form of ethnographic films, rather than interpreting the film, and focusing on the embedded meanings and significance of the film, interact with the text in different ways from those expected by the researcher. For Custen (1982: 240) this is manifested in what he called 'talk through film', that is where the viewers discuss how the film is meaningful to them and they find this meaning by drawing upon their experiences and lives separate from the film. Thus a distance is created between the research object (the film) and the individual. In addition, the technical nature and flow of moving images or films acts as a barrier to the interactive element of the interpretation process.

One of the practical approaches for the presentation of film-based visual images within printed texts is to identify particular images from the film that may be used to demonstrate a theme or sequence of events that have been recorded. Thus, the use of still images can almost create a storyboard to support the text and the research. A good example of this is where researchers may be filming a festival, and there are four or five significant aspects of the festival that are particularly important for the research. By isolating these from the original film in the form of stills it will be possible to show a progression of events or impacts. Although the reader may not have full access to the film resource, their interpretation and relationship to the research can still be enhanced by understanding the significance and sequencing through the excerpts from the empirical data.

Context

An important consideration in understanding how you present your visual data is to recognise the significance that the captions and titles you attribute to particular images or scenes have on the context and interpretation of the image. It can be argued that the text that surrounds visual images becomes a reactor, and a transactional reaction is formulated by the reader as the 'text directs perception' (Kress and van Leeuwen 2001: 67). This reaction is created by reinforcing 'signposts' that guide the reader in a particular direction. For example in a tourism brochure, the initial pages may show a deserted rugged cliff scene (text hidden) that may be geographically and culturally non-specific or generically anonymous. The picture could represent any number of non-specific destinations, however the text on the page or opposite page will direct the reader to the advertised destination providing context and direction. Thus, it can be argued that the text directs perception while guiding the significance of the landscape geographically, socially and culturally.

Thus, conceptually, a conversion process takes place in which meaning is guided by changing perception through various techniques such as the use of text and changes in context, Kress and van Leeuwen (2001: 67) call this process 'participant relay'. This relay denotes a text-image relationship in which the text extends or re-conceptualises the visual information. Furthermore, the interpretation of narrative images is also guided by the inclusion of what Kress and van Leeuwen define as 'secondary participants' (1996: 67). These participants are not directly related via vectors but become related in other ways or circumstances (Kress and van Leeuwen 2001: 71) i.e., the 'setting' of the narrative images, thus highlighting the nature both conceptually and narratively within the text. For example, the tourist represented within the brochure may create the vector, but the locals in the background highlight the nature of the relationship between both the host and guest and the position of the tourist within host/guest/cultural relationship. Therefore, captions and text headings tie the visual images into the text and as such provide meaning and significance, but you need to be careful how you use these and what impact they have on the interpretation process (Banks 2006).

CHAPTER SUMMARY

The use of visual images in tourism studies provides a significant source of research materials although it is often seen as problematic because of the issues relating to the presentation of visual data within publications. However, it is possible to overcome many of these issues through good planning, and more significantly, understanding the wider impacts of visual research methods and their relationship to ethics and copyright. This chapter has attempted to identify some of the major issues that can be summarised as follows:

- It is important to be aware that in many of the areas such as copyright there is advice and guidance readily available. As part of the research process researchers should ensure that their work is contextualised within a technical, legal and moral framework, and the regulations adopted by the researcher's academic institution.
- There are a wide range of secondary visual sources, and numerous primary methods for the collection of still and moving images. The presentation of

these in publication depends very much upon the purpose and approach taken. Consequently each research project needs to be assessed on its own merits and the target audience of the research.

- The use of any visual sources must be underpinned by good academic, ethical and legal principles and it is the researcher's responsibility to ensure that s/he is aware of developments within this area.
- Researchers should be aware of the technical requirements and guidelines for the presentation of visual materials in their chosen publication.
- Researchers must plan their research thoroughly as many of the issues related to the publication of visual materials can be overcome if they have been planned for. For example gaining copyright approval or informed consent in advance.
- Do not be discouraged from utilising visual approaches in tourism research.

Annotated further reading

Banks, M. (2006) *Visual Methods in Social Research*. London: Sage.

Marcus Banks offers an indispensible guide to undertaking visual research and draws examples from television and film, as well as photographs. He offers practical help in accessing archives and overcoming some of the practical issues facing researchers and students.

Kress, G. and van Leeuwen, T. (2001) *Multimodal Discourse: The Modes and Media of Contemporary Communication Discourse*. USA: Bloomsbury.

Kress and van Leeuwen offer an insight into the relationship between multimedia and multimodal design thinking and product thinking, charts how the communication works between producer and consumer, and explain the process of communication which as authors we enter into. The book also helps us understand how the reader relates to the text and what we have presented. Although adopting a multimodal approach it is still heavily orientated toward visual methods of communication.

Rose, G. (2001) *Visual Methodologies: An Introduction to the Interpretation of Visual Materials*. London: Sage.

Gillian Rose provides an insightful and thorough introduction on how to interpret visual materials. The book is approachable and offers a very good overview of many of the available methodologies. It is a useful book to utilise when planning and developing a research strategy, as it enables the researcher to identify the significance of the visual source within the research process and to assess how the chosen visual source can be presented in the completed work.

Selwyn. T. (ed.) (1996) *The Tourism Image: Myths and Myth Making in Tourism*. Sussex: John Wiley & Sons.

This edited collection examines how tourism images inform research debates within tourism studies and offers examples of ethnographic research in Nepal, analysis of secondary sources such as postcards and tourism brochures. Although the text is now dated it still develops some important research themes that are still relevant today, additionally the book also provides an exemplar of how to present and develop visual research for publication.

References

Banks, M. (2006) *Visual Methods in Social Research*. London: Sage.

Barthes, R. (1964) *The Rhetoric of the Image*. New York: Hill & Wang.

Botterill, D., Haven, C. and Gale, T. (2001) 'A survey of doctoral theses accepted by universities in the UK and Ireland for studies related to tourism, 1990–1999', *Tourist Studies*, 2: 283–311.

Custen, G. (1982) 'Talking about film', in S. Thomas (ed.) *Film/Culture: Explorations of Cinema in its Social Context*. Metuchen, NJ: Scarecrow Press.

Harper, D. (1998) 'An argument for visual sociology', in Prosser J. (ed.) *Image-Based Research: A Sourcebook for Qualitative Researchers*. London: Falmer Press.

Heath, S., Charles, V., Crow, G, and Wiles, R. (2007) 'Informed consent, gatekeepers and go-betweens: negotiating consent in child- and youth-oriented institutions', *British Educational Research Journal*, 33: 403–17.

Heisley, D. and Levy, S. (1991) 'Autodriving: A Photoelicitation Technique', *Journal of Consumer Research*, 18: 257–72.

JISC (2006) 'Roles and responsibilities for staff using images for teaching and research', retrieved on 17 February 2010 from www.jiscdigitalmedia.ac.uk.

Kress, G. and van Leeuwen. T. (1996) *Reading Images: Grammar of Visual Design*. London: Routledge.

Kress, G. and van Leeuwen. T. (2001) *Multimodal Discourse: The Modes and Media of Contemporary Communication Discourse*. USA: Bloomsbury.

Martinez, W. (1990) 'Critical studies and visual anthropology: aberrant vs. anticipated readings of ethnographic film', *CVA Review*, Spring: 34–47.

Mertens, D. and Ginsberg. P. (2009) *The Handbook of Social Research Ethics*. London: Sage.

Messaris, P. (1997) *Visual Persuasion: The Role of Images in Advertising*. London: Sage.

Porter, J. and McKee, H. (2009) 'Playing a good game: ethical issues for MMOG and virtual world researchers', *International Journal of Internet Research Ethics* 2: 5–37.

Prosser, J. (ed) (1998) *Image-based Research: A Sourcebook for Qualitative Researchers*. London: Falmer Press.

Riley, R. and Love, L. L. (2000) 'The state of qualitative tourism research', *Annals of Tourism Research*, 27: 164–87.

Rose, G. (2001) *Visual Methodologies: An Introduction to the Interpretation of Visual Materials*. London: Sage.

Schwartz, D. (1989) 'Visual ethnography: using photography in qualitative research', *Qualitative Sociology*, 12: 119–54.

Shay, H. (2009) *The Writer's Guide to Copyright, Contract and Law: How to enjoy Your Rights as an author and Comply with Your Obligations* (4th ed.). London: How to Books Limited.

Stokes, S. (2009) *Digital Copyright*. Oxford: Hart Publishing.

Wiles, R., Prosser, J., Bagnoli, T., Clark, A., Davies, K., Holland, S. and Renold. E. (2008) 'Visual ethics: thical issues in visual research', retrieved on 25 March 2010 from: www.eprints.ncrm.ac.uk/421/1/MethodsReviewPaperNCRM-011.pdf.

Wiles, R., Coffey, A., Robinson, J., and Prosser, J. (2010) 'Ethical Regulation and Visual Methods: Making visual research impossible or developing good practice', retrieved on 25 March 2010 from http://eprints.ncrm.ac.uk/421/1/Methods ReviewPaperNCRM-011.pdf.

Part 5

Conclusion

12 The future of visual research methods in tourism studies

Donna Chambers and Tijana Rakić

> The visual is sorely underdeveloped in modern scholarship. Art history has attained only a fraction of the conceptual sophistication of literary criticism. Drunk in self-love, criticism has hugely overestimated the centrality of language to Western culture. It has failed to see the electrifying sign language of images.
>
> Camille Paglia

Introduction

The quotation at the beginning of this chapter from Professor Camille Paglia, controversial author, social critic, avowed feminist and teacher (Birnbaum 2005) we would argue, is also true of the state of tourism studies which has for too long failed to recognise the integral role of the visual. Indeed, ten years ago Marcus Banks had argued that 'visual images are ubiquitous in the lives and work of those who study and those who are studied' (Banks 2001: 179). It was surprising therefore that especially in the context of inherently visual phenomena such as tourism, there had been such a dearth of engagement with the visual, although pursuant to the publication of John Urry's seminal text on the tourist gaze in 1990 the question of the visual took centre stage within tourism studies. Since then, and as the chapters in this text have undoubtedly testified, there has been increasing awareness in tourism scholarship of the value of the visual in enhancing our knowledge and understanding of tourism discourses and practices.

The question that now confronts us, as we conclude this introductory text on visual research methods in tourism studies, is what might the future of visual tourism scholarship hold? Admittedly it is difficult to predict what will happen in the future and it is not our intention to be prescriptive about visual methods or their future in tourism research in any way. Nevertheless it is still useful to make some informed commentary about how we perceive the future role of visual methods in tourism studies as we enter the second decade of the twenty-first century and as tourism and tourism's visual culture becomes increasingly ubiquitous and important for a range of stakeholders including the tourists, the host populations, the marketers and the managers of destinations, various tourism-related businesses, and also tourism scholars who are engaged in researching and studying tourism. So the aim of this chapter is to provide insights into what we believe the future might hold for visual research methods in tourism studies. In terms of structure, the chapter mirrors that of this volume to the extent that our discussion of the future for visual research methods in tourism focuses on epistemological considerations, methodologies and methods and finally examines issues associated with the publication of visual tourism research.

The future of visual methods in tourism research: questions of epistemology

Epistemologically, we suggest that issues of intersubjectivity, reflexivity and embodiment will become ever more important in visual tourism research as the integral role played by both the visual researcher and research participants in the co-creation of tourism knowledge is more widely acknowledged and celebrated. The concept of embodiment is of particular importance as Urry's (1990) privileging of the visual or the power of the gaze has been deconstructed and destabilised by subsequent tourism scholars (e.g. Veijola and Jokinen 1994) who argue that it is necessary to understand the role of the body or corporeality in understandings and experiences of tourism. Indeed it is now widely understood that tourism is about multi-sensory, embodied experiences and visual tourism researchers must be able to understand and interpret these experiences through their own actions and interactions with their research participants. These are the issues that concern Scarles in her discussion of visual autoethnography and participant-driven photo-elicitation in Chapter 5. In terms of participant-driven photo-elicitation, Scarles argues that this can empower research participants and allow both researchers and participants access to their emotional selves, thus enabling knowledge of the many and varied creative and innovative practices within the tourist experience. With regard to visual autoethnography, in her chapter Scarles explains that 'where photo-elicitation solicits a dynamic, co-constructive collaboration between respondent and researcher, within visual autoethnography the researcher becomes more deeply situated within the research as they themselves also become researched'. In Chapter 6, Cederholm continues the discussion of the method of photo-elicitation which she argues allows for participants' situated interpretations of photographs, and, in a similar fashion, Pocock *et al.* argue in Chapter 7 that video diaries can be used to access participants own realities and to obtain an emic or insider's perspective of the tourist experience. These chapters appear to substantiate our belief that in the future visual tourism research will be preoccupied with epistemological questions such as intersubjectivity, reflexivity and how to access embodied experiences as the limitations of adopting objectivist and ocularcentric approaches become increasingly recognised.

On the other hand it should be stated that these concepts are more consistent with constructivist or interpretive approaches to research but this is not to say that there will still not be a place for visual tourism researchers who adopt more objectivist approaches. Indeed, as indicated by Rakić in Chapter 2, there are many philosophical perspectives that can be adopted by visual researchers including positivist and mixed approaches which blur the boundaries between positivism and interpretivism. Perhaps we can speculate that the future of visual tourism research lies in the abandonment of a belief in the strict dualism between positivist and interpretive, between quantitative and qualitative, a dualism which some have argued is no longer relevant or necessary within social scientific debates. This is not a new argument because as far back as 1983, Richard Bernstein, who undertook an historical exegesis of the confluences and divergences between the two predominant philosophical paradigms (which he deemed objectivism and relativism), concluded that the similarities between these two ostensibly antithetical paradigms were more important than their differences. He argued that there was a need to move beyond a focus on the apparent dialectic between objectivism and relativism, between positivist and interpretive philosophies respectively and instead to interrogate the nature of the thinking that had led to the establishment of this dichotomous relationship.

The question for this concluding chapter is whether philosophical issues in visual tourism research will move beyond questions of objectivism versus subjectivism, relativism versus realism, quantitative versus qualitative, to a space in which there is some commensurability between the two approaches? It is difficult to conclude with any certainty on this issue but what we can say is that what we have witnessed in visual tourism research is what we consider to be an increasing preoccupation with more interpretive approaches which recognise the polysemic nature of visual images and which include concepts such as the aforementioned ones of reflexivity, intersubjectivity and embodiment. However, we propose that at the same time this will not mean a complete departure from the more positivist or objectivist epistemologies which draw on quantitative methods such as content and factor analyses and, in this vein, we will often see both quantitative and qualitative combined in one visual tourism research endeavour as, for example, within the research undertaken by William Cannon Hunter which incorporated the drawing methodology and which he discussed in Chapter 8 of this volume. Perhaps what we will witness is a more pragmatic approach to visual tourism research where particular philosophical approach(es) are adopted because of their usefulness in allowing access to specific knowledges, practices and experiences of tourism although the evidence thus far does not seem to point overwhelmingly in this direction.

Perhaps also visual tourism researchers will increasingly examine the more mundane and banal practices of tourism rather than the traditionally exotic and spectacular as suggested by Haldrup and Larsen in their examination of family photography in Chapter 9 and as Pocock *et al.* have done in Chapter 7 with their examination of tourists' video diaries. Indeed within the context of what Bauman (2000) referred to as 'liquid modernity' where it is difficult to keep abreast of the rapid pace of global and local change, Haldrup and Larsen suggest that it will be important to focus on the 'extraordinary ordinariness' of intimate social worlds.

The future of visual methods in tourism research: questions of methodology and methods

In the context of methodology we argue that visual methods in tourism research will need to draw on interdisciplinary perspectives if they are to benefit from the wider knowledges and understandings that can be brought to bear on tourism phenomena. Certainly in what is an increasingly complex, interconnected and dynamic global and local milieu and in which tourism has been adopted by almost all nations as an important economic, socio-cultural and political activity, visual tourism researchers will need to incorporate insights from a multiplicity of disciplines in order to gain a sound understanding of the discourses and practices of tourism. This is a theme recognised in all of the chapters in this volume and which was also previously acknowledged by Sarah Pink (2007) in her discussion of visual methodologies and demonstrated by van Leeuwen and Jewitt (2001) in their *Handbook of Visual Analysis*.

We suggest that some of the visual materials that are used in tourism research will continue to be obtained from traditional secondary sources such as printed brochures, photographs and postcards but that visual researchers in tourism will become more creative in terms of the types of secondary materials that are examined. For example in Jokela and Raento's discussion of secondary sources in Chapter 4 it was seen that even postage stamps provide viable sources of data for visual researchers. Other creative sources of secondary visual materials include paintings and artworks that depict tourism sites and sights as described in an article by Tribe (2008) which was briefly discussed in Chapter 3 by Chambers. However many of these secondary sources make use of still images and there is still a dearth of visual tourism research that relies on moving images as secondary data which is used for the purposes of analysis, although the work on film or movie induced tourism seems to be auguring changes in this context (e.g., see Beeton 2005, which even though not visual research per se draws on media and marketing perspectives to analyse film). On the other hand, in terms of primary research the heavy reliance on still images such as photographs as sources of visual data seems to be changing as seen in Rakić's discussion of her use of ethnographic filmmaking to explore the links between world heritage, tourism and national identity at the Athenian Acropolis in Chapter 2; the use of video diaries by Pocock *et al.* in Chapter 7; and William Cannon Hunter's use of research participants' drawings in Chapter 8.

It is evident that electronic secondary sources such as the Internet and digital media will become more dominant as printed visual materials decline in importance. Images of tourism (people and places) displayed and disseminated in still and moving image formats on the Internet (through official national tourism organisation websites, electronic (guide)books, brochures and social networking sites to name just a few examples) will increasingly serve as legitimate sources of visual materials for analysis (e.g., see the article by Yan and Santos 2009, briefly discussed by Chambers in Chapter 3 which used official promotional videos of China available from the Internet). While visual tourism research is still dominated by analyses of printed material, it is clear that with the ever increasing proliferation of new electronic media, visual researchers will no doubt make more use of these within their research projects.

The point in all this is that we envisage, and hope for, a future in which visual tourism researchers will become more creative in their choice of visual data drawing on a multiplicity of primary and secondary, still and moving images, materials which can be created for the purposes of a research project or are available from either printed or electronic secondary sources. That said, it is important to note that visual methods are likely to continue to be combined with traditional methods such as interviews, focus groups and participant observation. Visual methods, despite the technological advances we might see in the future, are still likely to be but one set of techniques among a range of methods used to access tourism knowledges, understandings and experiences. According to Banks (2001: 178), 'visual research methods are, or should be, a step along the way: a means to an end, not an end in themselves' and we propose that this will still be the case in the future of visual tourism research, notwithstanding the greater creativity that will no doubt be involved in the use of visual techniques.

As we emphasised in Chapter 1 and as the contributors reflected in their chapters, a theme that needs to play a central role in the planning, practising and teaching of visual methods in tourism are visual research ethics. While the question of research ethics is integral to any research project, this is especially so in visual research particularly due to the fact that, contrary to word- and number-based research, it might not always be possible or appropriate for visual researchers to anonymise their data by for example obscuring people's faces and places in still or moving images (e.g., see discussions in Prosser *et al.* 2008; Wiles *et al.* 2008). Therefore, there are a number of distinct ethical considerations that visual researchers will need to make and which are relevant in conducting research in an ethical and responsible manner. While it is of utmost importance that no harm is brought to bear on research participants and that their privacy and anonymity (if requested and deemed as needed) is assured and upheld, those issues 'which pose particular challenges in visual research [are]: informed consent, and anonymity and confidentiality' (Prosser *et al.* 2008: 11) and it is to these issues and ongoing debates in visual research ethics which visual tourism researchers will need to pay particular attention.

Given that it would have been an impossible task to address the wide variety of ethical considerations visual tourism researchers might need to make in any one of the chapters in this volume, as with any research, it is of utmost importance for visual tourism researchers to engage with visual research ethics, while simultaneously avoiding reactions of 'uncertainty ranging from individual anxiety to acute forms of group ethical hypochondria' (ibid.: 3). Above all, visual tourism researchers will need to address research ethics in their proposals and ensure that they have engaged in thoroughly informed ethical decision-making processes which have incorporated factors ranging from legal requirements, existing regulations, procedures and ethical committees to issues of informed consent, copyright, confidentiality and anonymity as well as their own moral principles (ibid.).

While, admittedly, as we have mentioned, ethical considerations traverse wider issues than those discussed in the various chapters of this volume, we argue that in the future, given the increasing global concern with ethics and responsible research practices, visual tourism researchers will necessarily be required to engage more thoroughly with the existing literature (see Further Reading in Chapter 1) and ongoing debates and developments in visual research ethics. They will also be required to explicate any relevant ethical issues in the publications resulting from their visual research and in so doing make their own contribution to the overall improvement of visual methodologies (see also discussions in ibid.).

Related concerns such as transparency and rigour in research processes and practices, as Chambers propounded in Chapter 3, are also integral to the future of visual research methods in tourism studies. These issues are important in primary research and are equally so in secondary research as Jokela and Raento discuss in Chapter 4 where they mention the need for visual tourism researchers who rely on secondary materials to undertake 'cross-checks' and argue for the value of inter-textuality. They also indicate that certain behavioural attributes are important for researchers such as respectfulness, considerateness and cultural sensitivity. No longer will visual tourism researchers be able to remain conspicuously silent on questions of ethics in their practices, presentations and publications of their visual research projects.

In terms of methods of analysis of visual data it seems clear to us that especially for qualitative visual tourism research, more sophisticated Computer-Assisted Qualitative Data Analysis Software (CAQDAS) packages will be developed which can handle large quantities of varied data including text, still and moving images. Some of these computer software packages were discussed by Sather-Wagstaff in Chapter 10 who also made very relevant and interesting comments on the advantages and disadvantages of manual versus computer-assisted qualitative visual data analysis. These comments, along with our own experience in qualitative visual data analysis, lead us to a conclusion that despite the ongoing development of the ever more sophisticated software packages, due to the advantages of manual analysis, such as the level of familiarity with the data being analysed which Sather-Wagstaff discussed in her chapter, some qualitative visual researchers will still continue to prefer relying on manual or a mixture of manual and computer-assisted qualitative visual data analysis techniques. On the other hand, visual tourism researchers who as a part of their projects rely on quantitative methods of visual data analysis are somewhat more likely to prefer relying extensively on computer-assisted analysis especially as the advancements in software progresses and researchers become ever more familiar with their use.

The future of visual methods in tourism research: issues of publication

Finally, with regard to the inclusion of visual research in tourism publications, Tresidder in Chapter 11 spoke of the limitations, opportunities and ethical considerations associated with incorporating still and moving images in this context. However as electronic publications become more ubiquitous and readily available, we can envisage a time when it will be possible to readily incorporate (audio)visual data and research outputs within tourism publications. In addition to the possibilities offered by the electronic format of academic publications where, for example, videos could feature as figures within electronically published journal articles (e.g., see Elsevier 2010 and their guidance for submitting videos which authors wish to embed within the body of the electronic versions of their articles), Pink (2007) provides an interesting discussion of the use of hypermedia representations in ethnographic work, which, admittedly has not yet taken off in tourism studies. An important consideration as Pink argues will be how these representations are 'consumed and appropriated in the practices and discourses of contemporary culture and society' (2007: 196). Indeed in the very act of consumption and appropriation new knowledges are created as is suggested by Pink who argues that hypermedia users:

> Produce knowledge by making sense of different types of information presented in multiple narratives and making the text coherent by producing their own linear narratives from it. Each individual user may follow a different route through the multiple narratives of a hypermedia representation, creating their own narratives and unique, experience-based knowledge, which will be inevitably informed by their wider biographical experiences.
>
> (Pink 2007: 197–8).

So, the question for visual tourism researchers will be not only how to include visual representations in their published materials but how this will be received and processed by consumers of this material and how in the act of consumption, new tourism knowledges are created. Given the rapid pace of change of technology it is perhaps inappropriate to go into further details about the ways in which visual materials might be included in publications because as soon as this book is published some of these discussions will no doubt become obsolete! Suffice to say at this point that we are optimistic for the future of publishing formats which are undeniably dependent on the creative and innovative thinking and decisions of editors, reviewers, authors and publishers of academic journals and books in tourism as they, along with the wider community of visual researchers, encounter dilemmas of adequate representation and inclusion of [audio-]visual data and research outputs in academic research publications. The increasing availability of electronic publications for both books and journals which can now be more readily adapted to accommodate the inclusion of still and moving images, along with the increasing acceptance of visual approaches to tourism research will probably play a central role in this process.

CHAPTER SUMMARY

In this chapter we have sought to portend a future for visual research methods in tourism studies although we fully recognise the challenges this involves particularly in the context of a rapidly changing global environment in which the clichéd expression that *'the only constant is change'* is perhaps the only viable consideration. Indeed we can say with certainty that the discourses and practices of visual tourism research will change but *how* they will change is still open to discussion. As mentioned in our introduction to this chapter we cannot be, nor do we wish to be, prescriptive about visual methods in tourism studies or what the future will hold. With these caveats in mind, we believe that it has nevertheless been pertinent to proffer some of our own thoughts on the future of visual research in tourism throughout this chapter and these are summarised below:

- Epistemologies of visual tourism research will increasingly include issues of intersubjectivity, reflexivity and embodiment as researchers come to recognise the importance of emic perspectives, the way in which knowledges about tourism are co-created by researchers and research participants, and the way in which tourism experiences are inherently multi-sensory and embodied.
- Methodologies of visual tourism research will increasingly involve inter-disciplinary perspectives as researchers strive to come to terms with tourism's visual culture, discourses and practices in the context of a continuously turbulent and dynamic global milieu. In this context researchers will need to thoroughly engage with visual research ethics as well as explicitly address these in not only their research proposals but also within the publications of their research findings.
- Visual research methods in tourism will become increasingly creative and diverse as researchers draw on multiple techniques in which electronic media will play a vital role. Analyses of visual materials will also become more sophisticated as technological advances are made in the software used for data analysis and as researchers become ever more familiar with their use.
- Visual tourism researchers will find it easier to publish their work without having to convert their visual research almost entirely into textual format as electronic publication outlets become more ubiquitous and as visual approaches to tourism research continue to flourish.

Annotated further reading

Banks, M. (2001) *Visual Methods in Social Research*, London: Sage.

This book by Banks is an important resource for any researcher seeking to undertake a visual research project whether in tourism or in the social sciences. Important for the discussions in this chapter on the future of visual research methods is Bank's extensive discourse on new technologies including image digitisation and computer based multi-media.

Grady, J. (2008) 'Visual research at the crossroads', *Forum: Qualitative Social Research*, 9: retrieved 30 May 2010 from www.qualitative-research.net/index.php/fqs/article/view/1173/2619.

This article engages in an interesting debate surrounding the current state of affairs of visual methods in the social sciences. It contains the claim that visual methods are currently at a crossroads from which they can either continue to be marginal or move towards becoming a more mainstream research activity.

Jokinen, E. and Veijola, S. (2003) 'Mountains and landscapes: towards embodied visualities', in D. Crouch and N. Lübbren (eds). *Visual Culture and Tourism*. Oxford: Berg.

This book chapter presents an autobiographical narrative discussion of how the authors have experienced tourist landscapes and includes an insightful account of how these experiences are embodied and often normalise extant power relationships in society.

Papson, S., Goldman, R., and Kersey, N. (2007) 'Website design: the precarious blend of narrative, aesthetics, and social theory', in G. Stanczak (ed.). *Visual Research Methods: Image, Society, and Representation*. London: Sage.

In this book chapter the authors present a lively discussion of the challenges and opportunities they encountered when they attempted to build a website to present their visual research data rather than using the traditional route of manuscript publication.

Pink, S. (2007) *Doing Visual Ethnography*. 2nd edn, London: Sage.

This text by Pink contains an interesting chapter on ethnographic hypermedia representation in which she includes a discussion of different areas of hypermedia practice namely on-line journal articles, on-line resources and CD-ROM and DVD publications.

References

Banks, M. (2001) *Visual Methods in Social Research*. London: Sage.

Bauman, Z. (2000) *Liquid Modernity*. Cambridge: Polity Press.

Beeton, S. (ed.) (2005) *Film Induced Tourism*. Clevedon: Channel View Publications.

Bernstein, R. J. (1983) *Beyond Objectivism and Relativism*. Oxford: Basil Blackwell.

Birnbaum, R. (2005) 'Camille Paglia interview', *The Morning News*, retrieved on 23 May 2010 from www.themorningnews.org/archives/birnbaum_v/camille_paglia.php.

Elsevier (2010) 'Author instructions: multimedia files', retrieved 18 May 2010, from www.elsevier.com/framework_authors/Artwork/Multimedia_2010.pdf.

Grady, J. (2008) 'Visual research at the crossroads', *Forum: Qualitative Social Research*, 9: retrieved 15 May 2010 from www.qualitative-research.net/index.php/fqs/article/view/1173/2619.

Pink, S. (2007) *Doing Visual Ethnography*. London: Sage.

Prosser, J., Clark, A., and Wiles, R. (2008) 'Visual research ethics at the crossroads', retrieved 15 May 2010 from www.eprints.ncrm.ac.uk/535/1/10–2008–11-realities-prosseretal.pdf.

Tribe, J. (2008) 'The art of tourism', *Annals of Tourism Research*, 35: 924–44.

Urry, J. (1990) *The Tourist Gaze: Leisure and Travel in Contemporary Societies*. London: Sage.

van Leeuwen, T. and Jewitt, C. (2001) *Handbook of Visual Analysis*. London: Sage.

Veijola, S., and Jokinen, E. (1994) 'The body in tourism', *Theory, Culture and Society*, 6: 125–51.

Wiles, R., Prosser, J., Bagnoli, A., Clark, A., Davies, K., Holland, S., *et al.* (2008) 'Visual ethics: ethical issues in visual research', retrieved 22 November 2009, from www.eprints.ncrm.ac.uk/421/1/MethodsReviewPaperNCRM-011.pdf.

Yan, G. and Santos, C. A. (2009) '"China forever": tourism discourse and self orientalism', *Annals of Tourism Research*, 36: 295–315.

Index

Lightning Source UK Ltd.
Milton Keynes UK
UKOW07n0054120115

244299UK00009B/142/P